GRAPHING STATISTICS & DATA

Anders Wallgren
Britt Wallgren
Rolf Persson
Ulf Jorner
Jan-Aage Haaland

GRAPHING STATISTICS & DATA

Creating Better Charts

SAGE Publications
International Educational and Professional Publisher
Thousand Oaks   London   New Delhi

For information address:

SAGE Publications, Inc.
2455 Teller Road
Newbury Park, California 91320
E-mail: order@sagepub.com

SAGE Publications Ltd.
6 Bonhill Street
London EC2A 4PU
United Kingdom

SAGE Publications India Pvt. Ltd.
M-32 Market
Greater Kailash I
New Delhi 110 048 India

Printed in Halmstad, Sweden

Library of Congress Cataloging-in-Publication Data

Main entry under title:

Statistikens bilder. English.
    Graphing statistics & data: Creating better charts / Anders
Wallgren . . . [et al.].
      p.   cm.
    Translated from Swedish.
    Includes bibliographical references and index.
    ISBN 0-7619-0598-7 (alk. paper).—ISBN 0-7619-0599-5 (pbk.:
alk. paper)
    1. Statistics—Graphic methods. I. Wallgren, Anders.
  HA31.S8313 1996
  519.5—dc20   96-7933

Graphic Design: Jan-Aage Haaland
Cover: Sage Publications, Inc.

This book printed on acid-free paper.

96 97 98 99 00 01 10 9 8 7 6 5 4 3 2 1

Why we wrote *Graphing Statistics & Data*

Society has become more and more dependent on statistics and other numerical information. However, all this numerical information is meaningless if it cannot be presented in a proper and easily accessible way. Charts and maps are effective aids to those who need to illustrate numerical data.

Our practical experience at Statistics Sweden, at universities and in business life shows how effectively charts can highlight the important points in a mass of data, illustrate complicated relationships and produce clarity.

At the same time experience shows that it is difficult to create good charts. The development of graphics programs for personal computers has, of course, made it technically simpler to create charts. However, programs have so many "refinements" that those who have not understood the basics of chart drawing have almost unlimited opportunities for producing unsuitable charts. The need for the chart designer to master the basic principles of creating good charts has therefore increased.

Recently statistical graphics have attracted increasing attention, not least in the U.S.A. This attention has also given new understanding of how readers perceive different types of charts.

To produce a good chart is both a skill and an art. As there was no book in Swedish which took into account both of these aspects we have written *Statistikens bilder*, the Swedish edition of *Graphing Statistics & Data*. We begin the book with a survey of the basics of drawing charts, such as choice of chart type, and how to use the building blocks of charts, such as axes, scales and patterns. Then we deal with the principles of creating effective and easy-to-read charts.

Throughout the book we have used real data as the basis of the maps and charts. In a separate chapter we show step by step how to work from the data to the finished chart in practical situations.

The book focuses on charts which shall appear in a report, in a book or at a presentation. Charts can and should be used as a working tool for you to get to know the data you are working with. The same principles apply for these working charts as for presentation charts, but, of course, they do not need to meet the same technical quality requirements.

This book is aimed at all those who have to illustrate numerical data – professional statisticians, analysts, marketing managers, economists, researchers, students, journalists and illustrators amongst others. *Graphing Statistics & Data* does not require any formal training in statistics, but it would be an advantage if the reader was familiar with the basic principles of statistics and had experienced the need to be able to create good charts.

We hope that those who use the book will find many tips and a lot of good advice about how to create good charts. Good charts do not draw attention to themselves for their own sake, but make the reader aware of the qualities of the statistical data. Good charts *are* information.

Why we translated *Graphing Statistics & Data*

When the Swedish edition of *Graphing Statistics & Data* was published we were contacted by colleagues who suggested an English translation. They pointed out that there was no compact description of the basics of drawing charts in English either. The fact that this interest came both from Swedes and foreigners and from both individual organizations and international agencies persuaded us to follow the advice.

We have kept most of the examples from the Swedish edition but some have been revised by using data from the U.S.A. To save space in the titles of the charts we do not explicitly mention Sweden in many Swedish examples. In some charts we use SEK, an abbreviation for Swedish crowns.

4

Contents

Reading instructions:

Chapters 4 to 10 deal with different types of charts. If you do not want to start by reading about all the types of charts in one go, you can leave some of these chapters until later. Regardless of what types of chart you want to work with, you should read the following chapters which deal with general concepts and principles:

Chapter 2 *Choosing chart types* and Chapter 3 *The building blocks of charts*
Chapter 11 *Some chart philosophy* and Chapter 13 *Charts in practice*

You should also practice drawing charts with your own data (some data can be found in the book). Use whatever graphics program is available to you. You can then try varying the form and building blocks of the chart for each set of data. Learn by comparing the different charts.

1 The power of charts

Charts speak directly to the reader

A set of data can be presented either in a table or in a chart. Tables have several advantages: the reader is given a compact and structured synthesis and many details can be shown in a small area, but the reader only sees figures. To understand the table's contents many figures have to be compared and evaluated.

"A chart says more than a thousand table cells" is statistic's version of the old proverb. Charts do not show exact details, but give an immediate depiction of the differences and patterns in the set of data. The reader can see immediately where there are major differences or similarities without having to compare and interpret the figures himself.

People charged with serious crimes per 100,000 inhabitants, 1842-1988

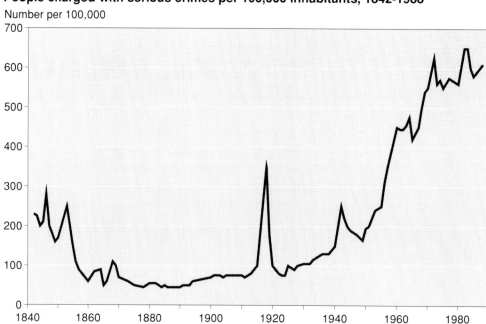

The chart above summarises Swedish criminal history for a period of almost 150 years. The reader quickly sees both the long-term trends and temporary deviations. We can see that serious crime declined during the mid eighteen hundreds only to rise again during the nineteen hundreds and reach a very high level in recent times. We also clearly see the large increases in connection with the two World Wars.

The eye can take in all this information quickly because the chart itself is very simple – a line chart with only one curve. Think how long it would take to read the corresponding figures in a table and then create for yourself a picture of the development.

At the same time we must remember that the chart above raises just as many questions as it answers. For example, what is "serious crime"? Has the definition of crime or the practice of the courts changed during the period in question? Is the number of people charged with serious crimes a good indicator of the level of crime? A discussion of the relevance and other qualities of the set of data becomes even more vital if it is illustrated by a powerful chart.

> Charts speak directly to the eye and are very effective in creating a picture in the reader's mind. Therefore the person who creates the chart has a special responsibility. In the same way that good charts convey information, bad charts convey disinformation. You have to create your charts in such a way that you do not mislead the reader.

The eye has a fantastic ability, but at the same time certain limitations. To draw good charts it is important to know how the eye works. In chapter 11 we will briefly touch on some of the results of what is generally called perception theory

Charts show both details and the whole

The eye is trained to see both broad outlines and small details. When we see a person for the first time we automatically register sex, approximate age, how the person is dressed etc. We also notice details, such as jewellery and distinguishing marks. Naturally, exactly what we notice varies from person to person and from situation to situation.

When we see a chart too we immediately register both the main features and certain details. In the same way that we study a person who interests us more closely, so we penetrate after a while into the details of the chart. As chart constructors one of our aims is to capture the reader's attention so that he or she will study the chart more closely.

Maps have a special place in showing spatial variations because they use two dimensions. A table or a bar chart can never be as effective at showing similarities and differences between adjacent areas. The same is also true of a time series chart. No table can give as good a picture of the progression of time as a line chart.

The two thematic maps to the right show the difference between the presidential elections in the USA in 1988 and 1992. We see immediately the large change in the relative strengths of the two parties. We also see how the democrats' regions in the north-west, north and north-east expanded from one election to the other.

With a more detailed study we can detect deviations from the general pattern and begin to ask ourselves why individual states voted the way they did.

Presidential elections in the USA in 1988 and 1992

1988

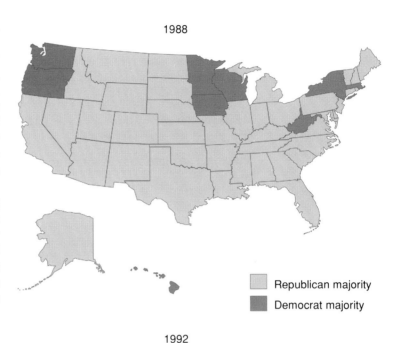

Republican majority
Democrat majority

1992

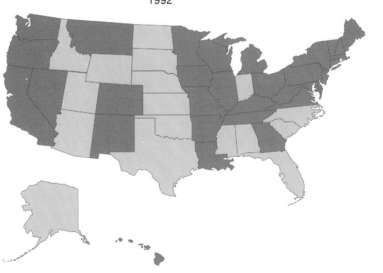

As chart constructors we must find a balance between details and the whole. The chart should not be overloaded so that the reader "does not see the wood for the trees". At the same time it is important to provide a lot of information in the chart. In this balancing act it is a good idea to remember that it is what the reader perceives which is important, not how much you manage to include as a constructor.

Charts encourage comparison and analysis

The eye is trained to recognise patterns. Originally this was in order to interpret man's natural environment, to see if there were tracks of game or signs of danger. Now it is even more important to quickly perceive and recognise different symbols. When we read text we do not read individual letters but immediately recognise words and perhaps even whole phrases. We also find it very easy to immediately distinguish between two persons, for example.

In a similar way the eye can register differences in the lengths of bars, gradients of lines etc. A chart stimulates the reader to look for patterns and connections. By presenting two or more charts together, we encourage comparisons.

Charts can be used in various roles in a publication. One role is to be a gateway which awakens the reader's desire to read on. Another is to develop ideas further, to illustrate and to explain. Different roles make different demands on the chart, but the common theme is that the reader should be encouraged to go deeper into the area or problem concerned.

On the previous page we placed two maps one above the other in order to encourage the eye to make comparisons. In the chart on the right we have placed many bars one above the other for the same reason

Intake by sex into three-year courses at upper secondary school, autumn 1991

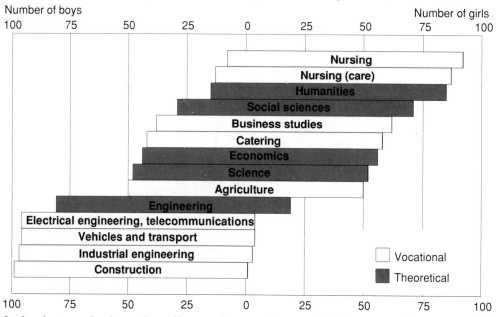

In the chart we clearly see the differences between boys and girls in terms of intake into upper secondary school. It shows forcefully how certain subjects are dominated by one sex. The girls dominate in nursing and in humanities etc. The boys dominate in technical subjects, with almost 100% boys studying construction.

Both the eye and the brain may need help to recognise patterns. In the chart above they are helped by the fact that the subjects are ranked from the one most dominated by girls to the one most dominated by boys. Different patterns have been used to show the difference between theoretical and vocational subjects.

There are many alternative ways of showing the data behind the chart on the right. We shall return to this data in chapter 13

Charts encourage comparison and analysis. But as the designer it is you who decides what you shall use as encouragement. The chart above could give a totally different impression if, for example, you include or exclude subjects with few students or combine the subjects in a different way. It is important to account for the choices that have been made and to make them in as honest a way as possible.

Charts – themes with many variations

A chart can have almost any kind of appearance. Many chart types are very specialised and might be used only in certain areas, such as rank-size diagrams in human geography. However, most charts are variations on a few basic themes. So, for example, the chart on the previous page is a horizontal bar chart, where the bars are displaced relative to the common scale.

In this book we pay most attention to the principles which are common for the majority of charts. The chart below may appear unique, but actually it is composed of simple components – an underlying map on which a flow chart has been superimposed, supplemented with a line chart and text.

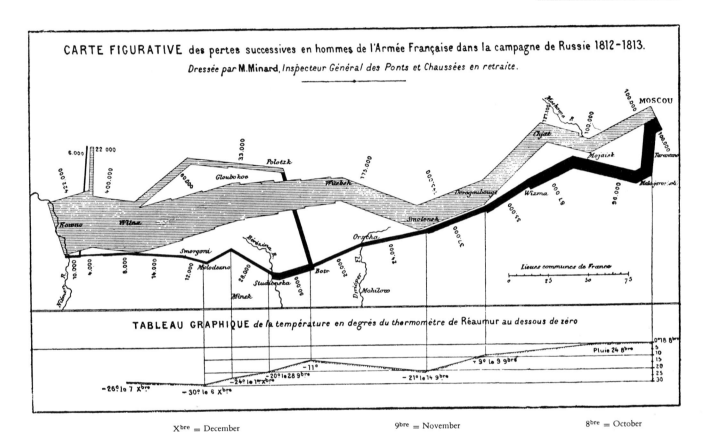

Xbre = December   9bre = November   8bre = October

This chart was drawn in 1861. At first sight it may appear that it is overloaded and that it is not even a chart. But look closer! It shows, superimposed on a map indicated by some place names and rivers, how La Grande Armée dwindled away when it marched into Russia. Of more than 400,000 men only a quarter reached Moscow. During the retreat (note the timescale and temperature scale underneath the map) the catastrophe was completed. Only 10,000 men returned.

The chart gives a fantastic summary of Napoleon's campaign. It is possible that Tolstoy did the same thing more artistically in War and Peace, but he needed a thousand pages to do so.

Edward R Tufte considers the chart of Napoleon's campaign to be perhaps the best diagram ever. We have been given permission to reproduce it from his book *The Visual Display of Quantitative Information*

Drawing charts is both a craft and an art. The craft involves knowing the basic charts which exist and how these should be drawn correctly from a technical viewpoint. It also involves an understanding of the principles of creating a good chart. How these basic elements are combined, how you choose the right chart in difficult situations, is something of an art. We attempt to cover both of these aspects in this book.

2 Choosing chart types

As is evident from the brief summary at the end of this chapter, there are a large number of different types of chart. How should we choose between these different types of chart? The answer to this question depends first of all on what problem we wish to illustrate or solve with the chart. However, there are no simple rules to tell us, for example, that the relationships between variables are best illustrated with a scatterplot or that a bar chart should be used for data which refers to a single point in time.

We need then to consider many different factors in choosing the best type of chart for a particular situation. Sometimes we have to choose between appropriate and inappropriate charts and sometimes we also have to choose between several types of charts which in themselves are correct. The first thing that we need to decide is what is the main thing we want from the chart: to show development over time, to show variations etc. This also has a natural link with the data itself.

In order to draw a good chart we must both emphasise the principal characteristics of the data and take into account any limitations these characteristics may present. The characteristics which are important in this context are the structure of the data, the relevant types of variables and what characteristics our measurements have.

The structure of the data

We first distinguish between whether we have measurements at one or several points in time. In the first case we speak of *cross-sectional data*. When we have data from a variety of points in time we speak of *time series data*.

No. of children	No. of families
0	15
1	21
2	12
3	6
4-	4
Tot	58

Cross-sectional data with one variable
In the table to the left we have data on 58 families. The data refers to a particular point in time and we have a single variable, i.e. the number of children. "Number of families" is not a variable, but the frequency for the various values (0,1,2 etc.) which the possible variable can adopt. Data with a single variable is also called univariate or one dimensional.

Height	Weight
52	4.5
49	3.8
48	2.9
50	3.4
49	3.7

Cross-sectional data with two variables
This table shows the height and weight for five newborn babies. Thus we have two measurements for each child. The fact that the measurements are put together in this way is an obvious requirement for us to be able to study the relationship between the variables. This data is known as bivariate or two dimensional.

No. of children	No. of families Town	Country
0	8	7
1	12	9
2	6	6
3	2	4
4-	2	2
Tot	30	28

Cross-sectional data with two categories
In this example we have divided the 58 families from the first set of data according to where they live. We now have two categories, urban and rural, while the variable being studied remains the number of children in the families. Data divided into categories with one variable may also be regarded as two dimensional data. The two variables in our example would then be the number of children and the place of residence.

Age (years)	Weight (kg)
1	8
2	11
3	15
4	17

Time series
This fourth set of data shows the weight of a child at four different points in time. It is therefore a (very short) time series. We can also see it as data with two variables, i.e. time and weight.

Multidimensional data also exists, but here we content ourselves with two dimensions

Different types of variable

The problem we are working with and the structure of our data give us a certain framework within which we choose appropriate chart types. If, for example, we want to demonstrate how urban and rural populations differ in terms of family size, we may choose a grouped bar chart. If, on the other hand, we had chosen to compare the same families according to the age of the father, then a grouped bar chart would not have been appropriate. We also need to take into account which variable we are illustrating.

Age, weight, number of children, place of residence etc. are examples of different variables which we may want to illustrate with a chart. Different types of variables require different types of chart. Even if, for example, the number of children and age are both usually indicated by whole numbers, they have completely different characteristics and, as we shall see, are illustrated by different types of chart (bar charts and histograms respectively). The diagram below summarises the division into different types of variables which we need to be able to choose the correct type of chart.

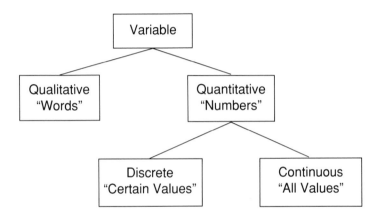

The first division is made according to how we characterise the value of the variable. A numerical variable, i.e. a variable for which we indicate the value using numbers, is called quantitative. If instead we use words then we speak of a qualitative variable. The value of the variable for place of residence is described in words (town-country) and is a qualitative variable. A variable such as weight (8 kg etc.) on the other hand is quantitative.

Among quantitative variables we distinguish between variables which can only take certain values and those which can take all the values in a range. Variables which can only take certain values (often whole numbers) are called discrete. Thus the number of children is a discrete variable. Variables which can take all of the numbers in a range are called continuous. Age is a good example of a continuous variable. The fact that we usually round off ages to whole years does not change this. It is the characteristics of the variable and not how it is shown which determines whether the variable is continuous or discrete.

Different types of charts suit different types of variable. A typical example is the histogram which is used for continuous variables. It would not be appropriate to use a histogram to illustrate a discrete variable.

In certain cases numerical values are given for the variable values of a qualitative variable. For example, "Very good" can be called 5, "Good" 4 etc. However, this does not alter the fundamental characteristics of the variable

The term scale type is also used in place of level of measurement

Levels of measurement

The variables we illustrate may have been measured in a variety of ways. In certain cases the type of variable directly dictates how the measurement is carried out, but in other cases there is a choice. The variable "age" can, for example, be given in years, but we can also have measurement values such as young – middle aged – old. Different types of measurement give different amounts of information about the variable concerned. This influences our choice of chart. We differentiate between four levels of measurement as shown in the table below.

Level of measurement	Characteristics. The Measurement Values can be ...
Nominal scale	Distinguished
Ordinal scale	Distinguished and ranked
Interval scale	Distinguished, ranked and measured with constant units of measurement
Ratio scale	Distinguished, ranked, measured with constant units of measurement and have a zero point

The nominal scale gets its name from the Latin "nomen" meaning name. The ordinal scale is so called because the order in which the values are placed has a significance. Correspondingly intervals and ratios have a meaningful significance in interval and ratio scales respectively.

The place of residence variable (town-country) is obviously measured on a nominal scale, since we are not able to rank these two measurement values in any meaningful way. Alphabetical order or, for example, order based on which value has the highest frequency are, of course, not considered to have any meaning in this context.

If we measure age according to the alternative young - middle aged - old we have an ordinal scale. We know which value is lowest or highest, but not how big a difference there is between the different values.

Dates are a good example of the interval scale. The zero point, the birth of Christ, is completely arbitrary and associated only with the western culture. However, the unit of measurement which is used, the year, is constant. In this context we can ignore the complications arising from the fact that a leap year has an extra day.

On the other hand, the measurement of time, such as the working week, is a ratio scale. If we were to observe the working time 0 for an individual this would mean of course that the person in question did not work at all. Correspondingly variables such as height, weight etc. are also ratio scales, even though we know that no child can weigh 0 kg. It is the characteristics of the scale which determine the level of measurement, not whether the values appear in a certain set of data or not.

Different levels of measurement give different charts

The level of measurement influences the choice of chart principally in that we ought to choose a chart which retains the characteristics of the measurement values. It is, for example, obvious that there ought to be an equal distance between the years in a time series chart and that as far as possible we should retain a zero point in order to be able to interpret ratios.

The level of measurement is also important in other areas of descriptive statistics. For example, in order for us to be able to calculate a mean value it is necessary that the variable we are working with is measured on at least the level of the interval scale.

A gallery of charts

There is an almost infinite number of different kinds of charts. However, most of them can be traced back to a limited number of basic types. In this book we shall consider the charts according to their applications – to show relationships, to show geographical variations etc. We shall start by introducing a kind of "Who's Who" for charts – a gallery of the basic types of chart. For each basic type we will give a very brief description of the main applications. You will also find a reference to the sections of the book where these types of charts are considered in greater detail. In these sections you will also find a number of variations on the basic themes.

Bar charts

Bar charts are perhaps the simplest form of chart. They are used in order to show numbers, proportions or other ratios. The variable which is described is qualitative or discrete.

Number of households by size of household 1985, in thousands

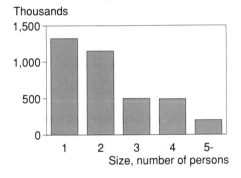

Horizontal bar charts

The bars can also be arranged horizontally. The range of application is the same as for the chart with vertical bars, but there is more room to write long names for the variable values.

Media habits of the population on an average day 1989

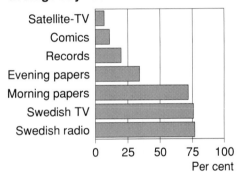

Grouped bar charts

Two or more categories may be compared in a grouped bar chart. Placing the categories alongside one another makes it easier for the eye to pick out the differences.

Families, by number of children, in the USA 1970 and 1990

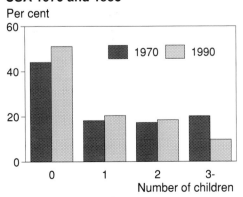

Stacked bar charts

A stacked bar chart gives a picture of how a total breaks down into its parts. We achieve this by placing the bars on top of one another.

Vacant apartments 1983-90 in private and public properties, in thousands

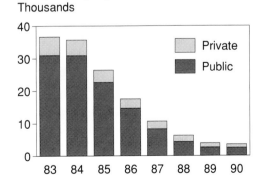

The charts in the book are based, with a few exceptions, on Swedish data

Vertical and horizontal bar charts are discussed on pages 24–25

Grouped and stacked bar charts are discussed on pages 26–27

Histograms

We use histograms in the same situations where we use bar charts, i.e. to show quantities and proportions. However, the variable to be illustrated is continuous. The freestanding bars are replaced with areas which are placed right next to one another.

Histograms are discussed on pages 30–31

Deaths in motor vehicle accidents 1988

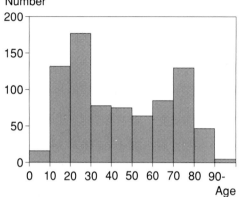

Population pyramids

In order to illustrate variations in gender and age we make use of population pyramids. A population pyramid consists of two "horizontal" histograms, one for men and one for women. One of the histograms has been reversed so that the two share the same scale.

Population pyramids are discussed on page 32

The Swedish population by sex and age 1987

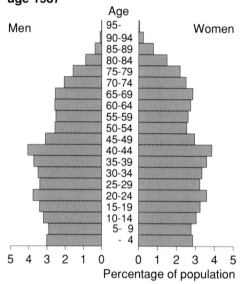

Pie charts

Pie charts are principally an alternative to bar charts. They are most appropriate for giving a general picture in situations where we want to compare proportions.

Pie charts are discussed on pages 34–35

Child care for pre-school children 1990

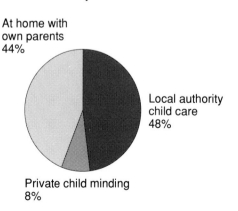

Scatterplots

Scatterplots are used to show how two variables co-vary (or how they do not co-vary).

Scatterplots are discussed on pages 46–47

Automobiles and personal income per capita, by state in the USA 1992

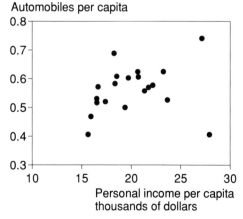

Line charts

Line charts are normally used for describing developments over time. Since time is continuous the different values are joined by lines. Line charts can also be an alternative to histograms.

Births and deaths in Sweden 1940-90

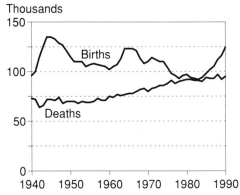

Area charts

Area charts are also used to show developments over time. Unlike a normal line chart an area chart shows how a whole is divided into components.

Number of unemployed 1970-90

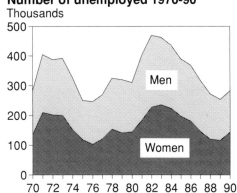

Line charts are discussed on pages 36–43

Area charts are discussed on pages 44–45

Flow charts are discussed on pages 54–56

Boxplots are discussed on pages 51–52

Flow charts

Flow charts show the flow between different states or between different regions etc. The flow is generally shown by arrows in such a way that the direction of the arrows symbolises the directions of the flow.

External trade of the USA, 1992

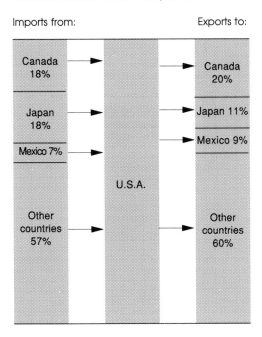

Boxplots

Boxplots are a relatively new type of chart which is used to show variations. By greatly compressing all of the data it is possible to show differences within a group and between groups at the same time.

Personal income per capita, by state in different regions of the USA, 1992

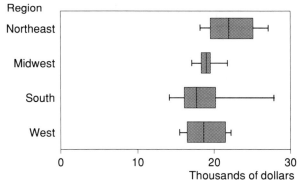

In this boxplot it is tempting to begin the x-axis at 10 thousand dollars. The chart would look more "interesting" without the empty space to the left. But the reader would receive an exaggerated image of the differences between the regions. For this reason we retain the zero point in this case.

We return to the question of whether or not the zero point should be included at several places in the book, e.g. on page 39

Density maps

Statistical maps are used to show geographical variations and relationships. In the density map below each dot symbolises a certain number of the thing we want to depict. Another common form of density map uses circles of different sizes to show variations.

Number of automobiles in different states in the USA, 1992

Density maps are
discussed on page 62

One dot corresponds to
1 million automobiles

Choropleth maps

In choropleth maps the different areas have different patterns to reflect the differences in ratios such as proportions, intensities, averages etc. The map below is also known as a hatch map.

Automobiles per capita in different states in the USA, 1992

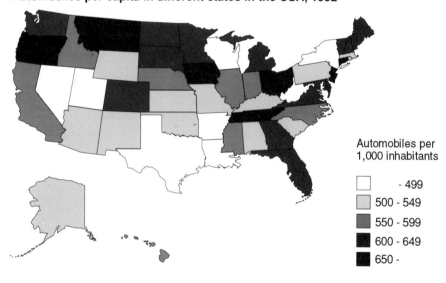

Choropleth maps are
discussed on pages 58–59

Automobiles per
1,000 inhabitants

- 499
500 - 549
550 - 599
600 - 649
650 -

3 The building blocks of charts

All charts are built up out of simple elements such as lines, areas and text. It is important to handle these building blocks in the right way so that the finished chart is easy to read and worth reading. Since the same principles on how to use the building blocks apply for all types of charts, we have chosen to deal with them in a separate chapter. Therefore, in the chapters on the various types of charts, we will only consider those aspects which are especially important for those particular charts.

Creating good charts is to a certain extent an art. As always with art a large measure of subjectivity is involved in what is considered to be a good or a bad chart. There are, however, certain established rules for designing good charts and we will cover these in this chapter. In chapter 11 we will return to the more fundamental ideas about the art of designing good charts. In the charts below you can see which building blocks we use and what we call them.

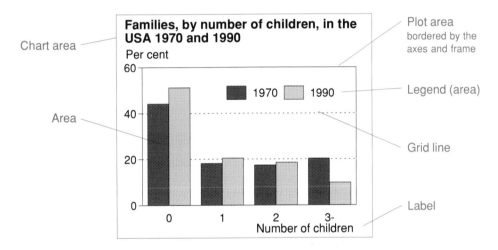

The chart area is an imaginary area which contains the whole of the chart, including the headings and all of the explanatory text. Normally the chart area is not marked out in any way, except when we put a different background behind the chart (see the next page).

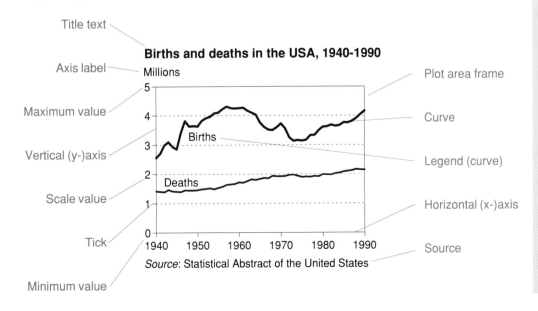

Sources should be given for all diagrams. If you are referring to your own research, then the source is clear and does not need to be given. In this book, where the diagram itself is the important thing, we have not given sources

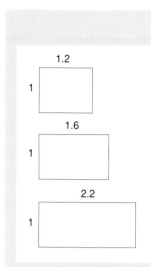

Plot area and chart area

The plot area is the area covered by the chart itself, without the titles, scale values etc. The plot area chosen must be big enough for whatever you are illustrating with the chart to be clearly visible. Charts with a lot of details need a larger plot area. The proportions of the plot area are determined to a certain extent by the data – a long time series requires a rectangular chart etc. As a rule a horizontal format is most suitable and if you are free to choose the proportions then the so-called golden section (about 1:1.6) is often a good choice. Sometimes the range 1.2 – 2.2 is given as an appropriate guideline for the ratio of the horizontal axis to the vertical.

Generally the chart area is considerably bigger than the plot area. It is the chart area which determines how much space the chart takes up when it is combined with text in a report or a book.

The plot area and the text together should fill the chart area in such a way that it will be perceived as a coherent unit. An effective way of achieving this is by making the chart "rectangular": write the title with a straight left-hand edge and begin it directly above the text which goes with the vertical axis. Split the title so that it approximately covers the chart and write the variable label below the horizontal axis. See the chart on the previous page.

Background

By placing a background behind the whole chart area you emphasise the rectangular shape of the chart and effectively bring it together to form a unit. A suitable way to create the background is with light shading so that it does not interfere with the chart. If you have access to colour then a lightly coloured background gives a neat impression. It also looks good to just put the background layer behind the plot area.

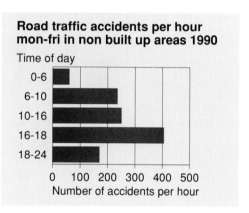

Framing and grid lines

Most charts benefit from framing, i.e. from the completion of the rectangle of which the vertical and horizontal axes form two sides. The chart is more coherent, thanks again to the rectangular shape.

Unsuitable charts

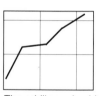

The gridlines should divide the plot area into areas of equal size

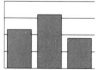

Charts with grid lines look strange if they are not framed

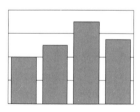
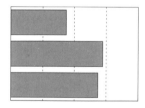
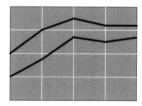

Grid lines make most charts easier to read, with the exception of very simple charts. They are placed at regular intervals on the axis where the eye needs help. We therefore use horizontal grid lines in vertical bar charts and vertical ones in horizontal bar charts, in both cases to make it easier to judge the length of the bars. In scatterplots and line charts it may be appropriate to use both horizontal and vertical grids.

Grid lines should be unobtrusive and should not distract attention from the data which is being illustrated. Make them as thin as possible. If you use a background then white grids are a good choice. You will find examples of how closely grid lines should be spaced on page 24.

Text in charts

Charts should be self-contained, i.e. all of the information necessary to understand the chart should be contained within the chart area. At the same time the chart itself should dominate, so the text should be kept to the absolute minimum necessary for the reader.

Titles and other text

Titles should describe in a concise way what is shown in the chart. It should be clear which group is being described (e.g. women, 18–64 years old), which variables are involved (e.g. cause of death), which year the data is referring to (e.g. 1991) and the type of data (e.g. percentages).

We consider it best to place the title above the chart in order to have a natural reading order. Using left-justification, i.e. writing the title in line with the left-hand edge of the chart, makes it easier to keep the chart area together in a neat way, see the upper chart in the margin. A title which has been centered, as in the lower chart, may, however, be used in situations where the chart will not be placed together with the running text, for example, when it is used as an OH picture.

The title may be written in a different style to the main body of text. An austere font, such as Helvetica, shows clearly where the main body of the text ends and the chart begins. Text in the chart, such as legends and labels, may be written using a somewhat smaller type than the title.

Make titles easy to read!

Titles themselves are more readable if you begin with the most important information. Thus:

Causes of death among women, 18–64 years old, 1991, in per cent

is much better than:

Percentage distribution in 1991 of the causes of death among women in the age group 18–64 years

Our recommendation

More information is given on different type faces and their uses on page 74

The main body of the text is usually written in Times Roman font with a size of 10 points.

The title should be placed above the chart and written in Helvetica, 10 or 9 points (depends on the size of the chart), bold

Text in charts should usually be written using Helvetica 9 points normal

Any sources should be written beneath the chart, in Helvetica 8 points normal, but with the word *source* itself in italics

---

All text in charts should be written horizontally!

We are used to reading text from left to right. The text in a chart should not be an exception to this rule. The text should be written horizontally, even if it is referring to a vertical axis or a sloping curve.

The reader should not, therefore, have to strain himself trying to read words where the letters are placed one beneath the other or words which are sloping at 45 degrees. Charts like the one on the right are an insult to the reader. A person trying to read a chart should not be forced to roll his or her head like an owl!

It is always possible to avoid sloping text in a chart. In order to create space for long explanatory texts, such as the names of countries, a horizontal bar chart may be used.

Unsuitable chart

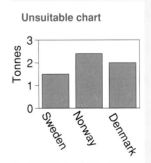

Axes, scales and ticks

The axes are made up of straight lines. There is no reason to use arrows. Text which relates to the axes is best placed at the top of the y-axis or on the right, below the x-axis. Obvious things such as "Years" in time series charts do not need to be stated.

The minimum value on the y-axis is often zero. The maximum value is chosen so that we are left with a little empty space above the highest values. It looks strange if bars or curves "touch the roof". We also make space for legends (see the following page).

If we have several charts close to one another then they ought to have the same scale. The eye can then make direct comparisons without being deceived.

Help the reader!

In order to help the eye we put ticks at regular intervals and give corresponding scale values. To avoid overloading the chart we can skip over a number of scale values and just use ticks. Usually one vertical and one horizontal scale will suffice. However, if, for example, you have a very long time series, you may need to repeat the scale to the right of the chart.

Scale values are chosen so that they refer to a "natural" numerical system. Standard practice is to build upon the numbers 1, 2 and 5. This means that the intervals on the scale are chosen as 1, 2 or 5, or alternatively 10, 20 or 50 etc. For decimal numbers 0.1, 0.2 or 0.5 etc. are used. Home-made scales of the type 30, 60, 90 etc. generally make the chart less easy to read.

An exception may be made with scales which show percentages. In this case the scale values 25, 50 and 75% seem natural. Another exception is when the scale values are of interest in themselves, such as the age of majority, which is 18 years in Sweden.

Remember the 1–2–5 principle!

Avoid big numbers on the scale. It is better to use thousands as a category on the axis and then write 100, 200 etc than to write 100,000, 200,000 etc.

Less suitable scales

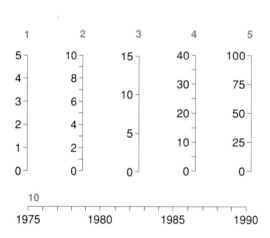

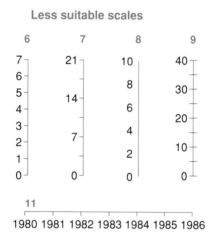

Scales 1-4 and 10 are examples of the 1-2-5 principle. Scale 5 is only suitable when it is referring to percentages.

Scale 5 is also an example of the fact that if scale values are of different lengths it looks neatest if they are right-aligned, i.e. make them finish at equal distances from the vertical axis.

It is important to bear in mind that the scale values should not be too close together. We think that the scales 1-5 and 10 have just the right number of scale values.

We recommend that you draw the ticks on the outside of the axes as we have done in the scales above.

Scale 6 and the time axis 11 have far too many scale values. Note that for scale 6 you would have to change the maximum to 8 in order to give every second value - otherwise you would come into conflict with the 1-2-5 rule.

Scale 7 is unsuitable because it uses a scale step of 3.5.

Scales 8 and 9 show that too little or too much spoils everything: as a rule you should use ticks, but do not let them dominate. For aesthetic reasons ticks should not be drawn across the axis, but only on the outside.

Legend (areas)

The same principles apply for areas and for curves. This means that we must always clearly state what each area or curve represents.

In charts with areas we can write the legend directly on the area, as, for example, in the area chart on page 15. Often, however, the areas are narrow and then we write the legend outside the plot area. In this case we put the text directly next to respective area, as in the chart on the right.

Sales of alcoholic drinks 1983-1989 in litres per inhabitant over 15 years

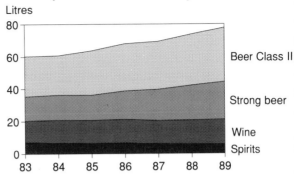

The variable 'type of drink' follows an ordinal scale (see page 12). We use this order in the chart

In both grouped and stacked bar charts we need to explain what the different bars represent. Legends can seldom be written in or next to the bars so we put a separate legend inside, next to or below the plot area.

Areas in legends should be put in the same order as they occur in the chart, from left to right, or from top to bottom. In a grouped bar chart this means putting the legend horizontally, either within the plot area as in chart A or beneath it as in chart B. In a stacked bar chart the legend is placed vertically – either within the plot area or to the right of it as in chart C.

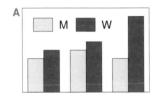

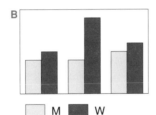

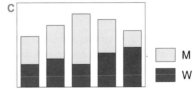

If the same variables reappear in several charts they should have the same pattern in all of the charts. The order of the variables should also be the same

Legend (curves)

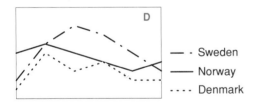

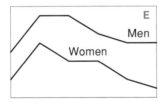

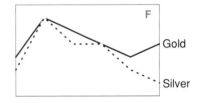

When we have several curves in one chart we need to make clear what the different lines represent. The legend can be arranged in different ways. Perhaps the most common is to put it beneath or to the right of the plot area (see chart D). This can give a somewhat uneasy impression since the eye is forced to jump between the chart and the legend. If possible, therefore, we try to put the legend next to the curves, as in E and F.

When there are only a few curves and they are well separated the legend may be written directly next to the respective curve. In chart E there is hardly any doubt as to which curve refers to men and which refers to women.

In chart F it is better to put the text outside of the plot area, but directly following the respective curves. Particularly with more complicated charts it may be advantageous not to burden the plot area unnecessarily.

With relatively big charts we can also put a conventional legend, such as that in chart D, inside the plot area.

We will return with more examples of handling explanatory text in time series charts in chapter 6

Patterns, shading and colour

We recommend that you always draw bars which are filled. To distinguish categories in a grouped bar chart, for example, you use different patterns for the different categories.

Usually you can choose between using patterns and shading. Normally shading, which is what we have chosen for most of the charts in this book, produces more restful charts. If you have the technical and financial means then colour is an alternative – used in moderation.

Area patterns

It is very important that you choose good patterns for your charts. Poor patterns can ruin the chart and also deceive the eye. For this reason, choose restful patterns.

In choosing between patterns and shading we prefer shading when we do not have too many different categories. Shading is actually nothing more than a very simple pattern. If we have a lot of categories or want to attract special attention then we use patterns.

Good area patterns

Dots and lines produce good and restful patterns. Lines need to be slanted as in the examples below. When you have several categories it is important to choose patterns which are clearly differentiated. If there is an order between the categories the patterns should reflect this order.

These patterns are relatively restful. We have placed them in the order which we consider natural. Avoid using more than 4 or 5 patterns.

Always put the darkest patterns furthest down. White is best put last in a pie chart.

Poor area patterns and poor combinations

It is not hard to give examples of poor patterns. Most PC programs have an abundance of patterns, many of which are so restless that the reader is almost physically repelled.

Unsuitable patterns or combinations of patterns

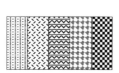

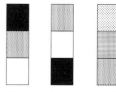

Awful patterns, not least when they are put next to one another.

Think of how different patterns interact. Avoid slanting in different directions. Also avoid horizontal lines (bars look shorter) and vertical lines (bars look longer).

If the darkest pattern is put at the top, the bar "collapses". Lighter patterns in the middle also produce a strange effect. And remember that the patterns should not be too similar.

Shading

Most of what has been said about patterns is also true of shading. Different shades must be easily distinguishable and arranged from light to dark. The four patterns on the left produce good distinctions. They can be complemented with white, and, in exceptional cases, black.

20%   40%   60%   80%

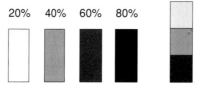

Do not use shades which are so similar that the eye has difficulty distinguishing between them!

Line patterns

If you only have one curve in a chart, for example a single time series, then you obviously make this an unbroken black line. The line should be sufficiently wide to be distinguished from the grid lines, but not too thick. A width of 0.5 mm is a good guide, while the grid lines may be 0.1 mm.

The line width should be adapted to the size of the diagram

If we have several sets of data in the same chart, we need to distinguish them from one another using different patterns. In theory we have access to a large number of different combinations of dashes and dots. In practice it is difficult to use more than 4 or 5, see the examples above.

We can also use different line widths. Again the choice is limited, since the thinnest line must be distinguishable from the grid lines and the thickest must not be too thick.

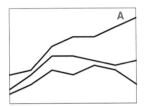 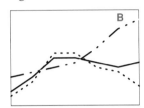 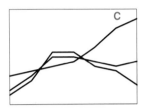

You must always bear in mind how the chart will look in printed or copied form. The technical equipment which you have access to limits your choice of patterns, shadings and line widths

If the curves never cross one another then you do not need to have different patterns, see chart A. However, for lines which intersect one another, as in chart B, it is necessary to use different patterns. If we draw lines as in chart C, the reader is forced to guess which sections of the lines go together.

Symbols

A time series consists of observations which are joined together by lines. You may choose to mark the separate observations (years, months etc.) with special symbols. On the right are three examples of lines with different symbols.

In our opinion symbols rarely bring anything other than confusion to a chart. They are seldom necessary in short time series and become too dominant in long ones.

On the other hand, symbols do have their uses in scatterplots and on maps. In both cases one should choose simple symbols such as circles or squares. They can be filled or empty and should be relatively small.

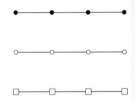

Colour in charts

In charts colour should be used sparingly! For charts printed in four colours with red, yellow, green and blue bars, the colours tend to obliterate the message of the chart. The costs of multicolour printing are also relatively high. If you have the opportunity, though, the use of *one* colour can be very effective. The charts below are examples of how you can make bars and lines stand out with the help of colour. Using a coloured background (perhaps with white grid lines) is also effective. You can then choose between dark bars against a light background or, as in the middle below, light bars against a dark background.

It is best to choose a discreet colour, perhaps blue or green, and to use relatively pale shading

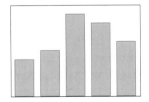 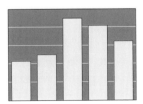 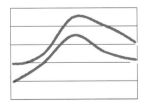

4 Bar charts – our basic chart

The bar chart is a kind of basic chart – simple both to draw and to read. It is used when we want to illustrate variable values which are distinct, i.e. with qualitative or discrete variables. Bar charts use spaces between the bars to emphasise this.

We shall begin by considering bar charts in situations where we want to show frequencies, but they are just as useful for showing sums and averages (see page 28).

Bar charts are used both for absolute and for relative frequencies. The appearance of the chart as such is not affected, but it is, of course, important to state what type of frequency is being shown. The charts below show the same data in these two ways. The chart to the left provides more information, namely the number of households, but the chart on the right may still be preferable in many situations. Percentages do, after all, make it easier to compare data sets of different sizes.

Occupants	Households
1	1,324,766
2	1,150,976
3	498,189
4	493,408
5-	203,001
Total	3,670,340

Number of households by size of household 1985

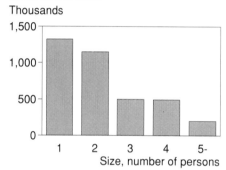

Percentage distribution of households by size of household 1985

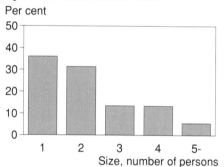

Gaps between bars

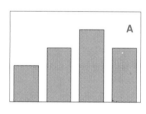
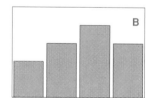
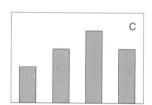

The bars should be considerably wider than the gaps between them, roughly as shown in chart A. However, there should be a distinct gap between the bars. Chart B with its narrow gaps looks too much like a histogram and is therefore unsuitable. The large gaps in chart C, on the other hand, make it look far too desolate.

Gridlines

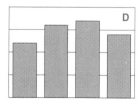
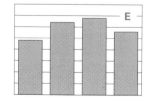
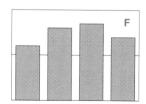

Grid lines in a bar chart are intended to help the eye to make comparisons and to read approximate values, as in chart D. Having grid lines too close together, as in chart E, gives a restless impression. On the other hand, the single grid line in chart F serves no purpose.

Horizontal bar charts

Normally we draw bar charts with vertical bars, i.e. with the variable to be illustrated on the x-axis and the frequency on the y-axis. However, there is nothing to stop us from swapping over the axes and putting the frequency on the horizontal instead. We then speak of a horizontal bar chart. Horizontal bar charts are to be preferred to vertical in two situations:

Variable values with long names

When the names of the different variable values are long, as are the descriptions of the occupations in the chart alongside, there is not enough space to write them under each value in a vertical bar chart. Since we should avoid slanting text (see page 19), a chart with horizontal bars is the only solution.

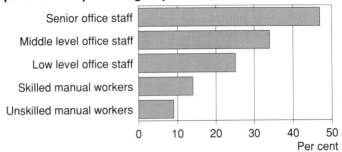

Proportion of people with higher education according to parents occupational group 1987

Many variable values

When we have many variable values, as in the chart on the right, we do not have space for all the bars and their accompanying text. This is true even if the names are shorter than in our example. With as few as between six and eight variable values it starts to get difficult to construct a neat and legible vertical chart. However, in a horizontal chart you can easily make space for the names of all European countries, for example.

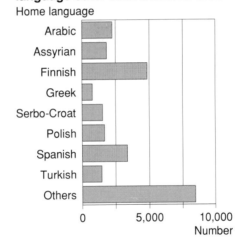

Pre-school children with a home language other than Swedish 1988

Right-justified text is the neatest for a horizontal bar chart. In this way you create a good connection between the explanatory text and the picture

Orders of the bars

The chart of home languages would perhaps have benefitted from being sorted into order of size. However, if this is done the category "others" should still be put at the bottom. It is after all a total of many of languages, each of which is less common than the language (Greek) which would come at the bottom if the list were sorted

Remember that a well-chosen order for the variable values gives a better chart! When you have a qualitative variable you have some freedom to choose the order. Arranging the values in order of frequency, from the most common to the least common, arouses interest.

An unsorted chart, such as chart G, gives a restless impression. Both of the other charts give a restful impression, but beginning with the largest category (chart H) creates a more "lively" impression than beginning with the smallest (chart J).

Grouped bar charts

Grouped bar charts are used for describing two or more categories at the same time. The different categories are represented by different bars in the same chart with common axes. In order to distinguish the categories we use different patterns or shading and a legend.

A grouped bar chart is suitable for showing two or three categories. If we try to cram in too many categories the chart becomes difficult to understand.

For the same reason we cannot show as many variable values as in a non-grouped chart.

It is difficult to give particular limits for the number of categories and variable values. As so often in the drawing of charts it is best to experiment. In situations with many groups it is often preferable to draw several ordinary bar charts instead.

Fishing opportunities for owners and non-owners of holiday cottages 1990

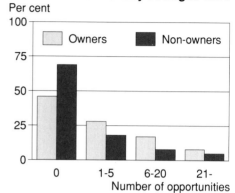

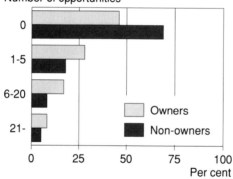

Naturally, a grouped bar chart may also be drawn using horizontal bars. The chart on the left shows the same information as the chart above. Notice that the two charts produce different effects, even though the plot area is the same size in both cases. In the horizontal bar chart the percentage axis has become longer so that the differences are seen more clearly.

Overlapping bars

By drawing the bars with an overlap we create a variation on the grouped bar chart. Overlapping bars save on space and the chart may appear more interesting graphically. Unfortunately there is also a risk that the overlapping will make the chart difficult to understand or misleading.

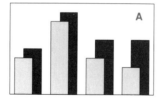
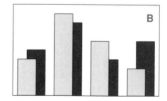
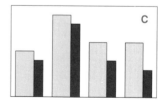

It is difficult to draw an overlapping bar chart with more than two groups

Overlapping is most suitable in situations such as that in chart A where one category has shorter bars throughout. In chart B, on the other hand, the overlapping bars create a restless impression.

It is best to put the category with the shortest bars in front with a light pattern, as in chart A. If instead the category with the shortest bars is put at the back, it will be obscured and may be perceived as being less important, see chart C.

Stacked bar charts

We use stacked bar charts in the same sorts of situation where we use grouped bar charts, i.e. in order to illustrate data sets consisting of two or more categories. In stacked bar charts, though, we put the bars on top of each other. For each variable value the height of the bar corresponds to the total frequency for all of the categories. Different patterns or shading show how the total is divided into different categories.

Similar limitations apply to stacked bar charts as to grouped bar charts. For example, it is very difficult to show a lot of categories and a lot of variable values in the same chart.

In stacked bar charts only the size of the bottom category is easy to read precisely. The reader only gets an approximate idea of the size of the other categories.

In the chart on the right we can see quite easily that in 1987 there were around 8,000 farms with less than 10 cows, but how many medium-sized farms were there? Correspondingly it is easy to see that the number of small farms has decreased, but it is very difficult to see what has happened to the large farms.

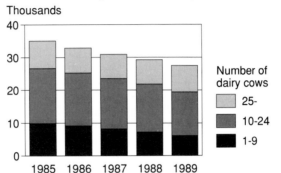

Farms with dairy cows 1985-89, in thousands

In both grouped and stacked bar charts we retain the order found among the variable values, for example, from small to large farms

The choice between grouped and stacked bar charts

Since both grouped and stacked bar charts are used in similar situations, we often have a choice between them. Our choice is determined by what we want to emphasise most.

In a grouped chart it is easy to compare the different categories with one another, but more difficult to form an understanding of the total of all of the categories. In a stacked chart, on the other hand, the total is clearly visible, whilst the size of the individual categories takes second place. In the charts below the same information is shown from two different points of view using these two different types of chart.

In the stacked bar chart on the left it is easy to see that the number of people doing independent work is roughly twice the number of those with skilled work. In the grouped diagram it is difficult to see this, but it is easier to see, for example, how women are divided between different levels of occupation. In the stacked diagram you do not even see whether there are any women in managerial positions.

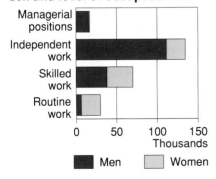

Number of industrial employees by sex and level of occupation 1990

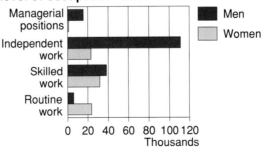

Number of industrial employees by sex and level of occupation 1990

Charts showing percentage distributions

A normal stacked bar chart shows how a total can be divided into its components. If we are interested in relative frequencies we use stacked bar charts where the totals are 100%. In these charts it is easy to compare the composition of different groups, but, of course, they do not give a picture of the size of the groups.

In many situations we have to decide how we are going to calculate percentages. The table below, which could be from an opinion poll in connection with the 1992 election in the U.S.A., provides an example:

Party identification Level of education	Democrat	Republican	Others	Total
Grade school	301	102	60	463
High school	751	459	180	1,390
College	208	209	46	463
Total	1,260	770	286	2,316

If we want to study how election behaviour varies according to the educational background of the voter, then we calculate the percentage horizontally. Each level of education is made to add up to 100% and we get the chart below on the left. If, on the other hand, we want to compare the political blocs regarding educational background of the voters, then we calculate the percentages vertically and get the chart on the right.

Different educational backgrounds give different voting patterns

Voters for different parties have different educational backgrounds

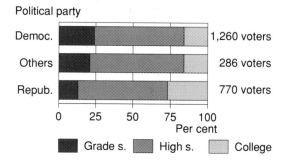

Since charts showing percentage distributions do not show the size of the respective groups, we should add this information. This can be done, for example, as in the chart on the right

Charts with averages or totals

Bar charts are normally used to show frequencies. As we have already said, bar charts can also be used to show totals, averages, ratios etc. These are put on the vertical axis in place of the frequency. It is often best to also give the actual size in the heading. In the chart below we use bar charts to compare the manufacturing industries in four countries.

Manufacturing industry in Denmark, Finland, Norway and Sweden 1988

It is very important to state clearly what the y-axis is showing: absolute or relative frequencies, totals, averages or something else

Number of companies in thousands

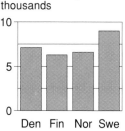

Number of employees in thousands

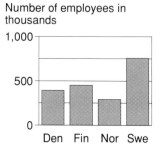

Number of employees per company

Negative values in bar charts

When we use bar charts to show frequencies or other positive values it is obvious that we only use the positive part of the y-axis. When the data consists of negative values or both postive and negative values we must also use the negative part of the y-axis. The charts below show growth rate and balance of trade which may be positive or negative.

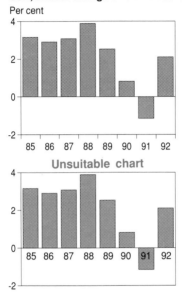

USA, annual change of GDP in constant dollars

Per cent

Unsuitable chart

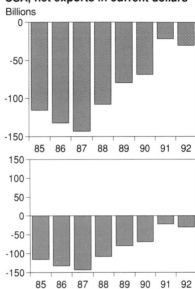

USA, net exports in current dollars

Billions

The scale values of the x-axis should always be written under the plot area, not under the zero line. The zero line should be clearly indicated.

If all values are negative you might skip the positive part of the y-axis. But if you want to stress that net exports may be positive you should include the positive part of the y-axis.

In spite of the fact that time is a continuous variable it is quite normal to use bar charts to illustrate development over time. Here we have chosen bar charts to indicate that each year is seen as a closed period.
If we have a long time series then a line chart is preferable (see page 36)

Dot charts

On page 24 we recommended the use of relatively wide bars in bar charts. This is in accordance with the principle of allowing the data to dominate the chart. In certain situations, especially with horizontal charts with a relatively large number of variable values, a dot chart may be a good alternative. In this type of chart the frequency value is indicated by a line which ends with a small, filled circle.

Per cent with higher education by occupational group of the parents 1987

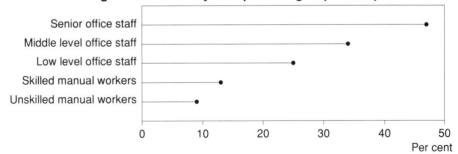

Per cent

This dot chart shows the same information as the horizontal bar chart at the top of page 25. Decide for yourself which you prefer!

Points to remember with bar charts

The quatities in a bar chart has a zero point which should be included in the chart. The y-axis should begin at zero, so that the bars give a true picture of the differences in levels. In other cases, such as charts which show development over time or relationships, we can make an exception from this rule.

Unsuitable chart

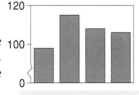

5 Showing frequencies

Bar charts with their different variants are only one way of showing frequencies. When we have a continuous variable we use histograms or frequency polygons instead. When we want to focus attention on proportions of qualitative variables we often use pie charts. This chapter concerns itself mainly with these types of charts.

Histograms

A histogram is used when we want to show frequencies for a continuous variable. The continuous variable can, of course, assume all values within an interval and the histogram reflects this by covering the whole of the interval concerned.

The starting point for a histogram is tables like those on the left. Histograms consist of rectangles with their respective classes as the base. In principle the area (= base times height) is proportional to the frequency. Like bar charts, histograms can be designed to show both absolute and relative frequencies (numbers or percentages).

Histogram comes from the Greek 'histos' which means area

Chapter 8 looks at charts which show variations. There you will find some alternatives to histograms

A

Age (years)	No.
0	555
1	528
2	397
3	404
4	357
.	...
.	...
18	309
19	296
Total	6,022

Children and teenagers in city X, 1990

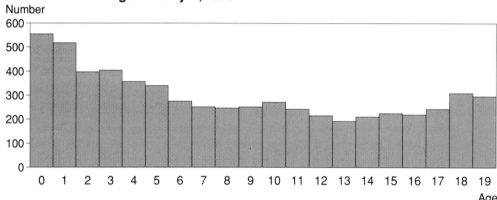

The variable 'age' in table A is continuous and is therefore illustrated with a histogram. When this is being drawn we have to remember that "1 year" means that the person has reached their first birthday, but not yet their second birthday. Since all of the data classes are one year long, the heights are proportional to the number of children in the data classes.

B

Age (years)	No.
0 - 4	2,231
5 - 9	1,368
10 - 14	1,132
15 - 19	1,291
Total	6,022

On the x-axis of a histogram we give either the class boundaries or the name of the class. If there is enough space then the name of the class is the most informative (see the two charts on this page). In the histogram at the top of the next page, however, we have used the class boundaries

Classes of equal size

Highly detailed tables, such as a complete table A, sometimes become too bulky and boring. For this reason we often compress the initial table by using larger classes. Tables B and C are two examples of this.

When we have classes of equal size, as in tables A and B, the base of each of the rectangles is the same. Therefore both the area and the height are proportional to the frequency. This means that we can show the quantities on the y-axis in the same way as in a bar chart.

Children and teenagers in city X, 1990

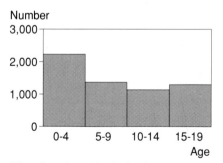

The data in table B is shown in this chart according to the same principles as for table A.

Since it is relatively complicated both to draw and to read histograms with classes of different size we recommend that, as far as possible, both tables and charts should be made with classes of equal length.

Different sized classes

When we have classes of different lengths we cannot make the height of the rectangles proportional to the frequency. Instead we have to calculate the height for each rectangle starting from its base, i.e. the width of the class. If one class is twice as long as another, then we have to divide its frequency by two in order to get the correct height.

When the classes are of different lengths, only the area is proportional to the frequency. We therefore write nothing on the y-axis, except perhaps for ticks.

Open-ended classes

In certain tables there are open-ended classes, i.e. classes which do not have a beginning or an end. We do not then know how long the class is and so we do not know how to draw the corresponding rectangle. Unfortunately, there is no simple solution to this problem. You have to do the best you can from case to case.

Sometimes you can treat an open-ended class as if it was closed. In a histogram concerning the ages of women having their first babies, for example, drawing the class "over 40" as "40–44" would not cause any great problems.

In the chart to the right it is difficult to decide on an upper limit in such a way. One solution is to end the histogram at 40 MWh and then put "over 40" as a rectangle a small distance away from it. Here we have chosen to state the percentage on the y-axis despite the fact that we have an open-ended class. In our opinion the reader will hardly be mislead either by this or by the treatment of the open-ended class.

The chart below demonstrates another solution. The chart ends with a straight line to indicate the open-ended class. This shows that the chart is open-ended, but the height of the final line is entirely arbitrary.

Pre-school children, school children and teenagers in city X, 1990

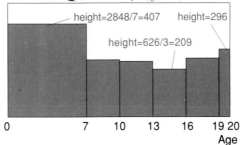

To show the data in table C you need to calculate the number of children per one-year class and give the rectangles the corresponding heights.

Consumption of electricity for heating of detached houses in 1988

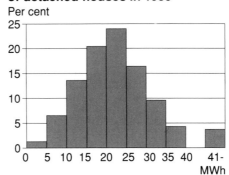

C	Age (years)	No.
	0 - 6	2,848
	7 - 9	751
	10 - 12	730
	13 - 15	626
	16 - 18	771
	19	296
	Total	6,022

In table C and the corresponding diagram we are justified in using classes of different length and class boundaries which deviate from the norm. The age of 7 is interesting as the age at which children start school in Sweden, and the others because they describe the stages into which school is divided

Fulltime farms according to income from agricultural procedure sold in 1989

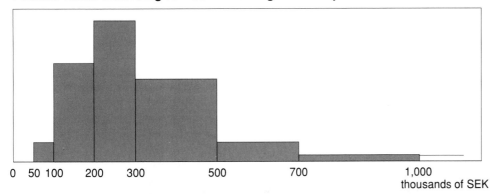

The chart on the left is an example of another situation where one is often forced to use different sized classes, namely very skewed distributions

SEK = Swedish crowns

Population pyramids are also called age-sex pyramids

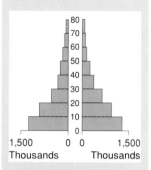

Anyone who wonders about the term "pyramid" can reflect on the pyramid for Zimbabwe above. The population pyramids for Sweden and other countries in Europe looked more or less like this when the term "population pyramid" was first introduced

Population pyramids

Histograms are almost always drawn with vertical rectangles. The reason for this is that the conditions which persuade us to choose horizontal bar charts, such as long names of variable values, seldom exist with continuous variables.

There is one exception, namely population pyramids. We use population pyramids to show the population of a country or an area divided according to gender and age. In the gallery of diagrams on page 14 there is a current population pyramid for Sweden.

A population pyramid consists of two horizontal histograms, one for men and one for women. One of the histograms is reversed so that the eye is able to make comparisons between the sexes. For all else the principles of the histogram apply; that you should avoid classes of different sizes and that you may have problems with open-ended classes. Here too you can choose between numbers and percentages. The percentages can either be percentages of men and women separately or percentages of the total population; the diagram should make clear which is intended.

Swedish residents born abroad by sex, age and marital status December 31, 1985

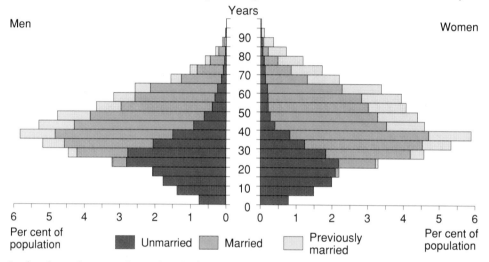

In the chart above we have divided the rectangles in the population pyramid into sections according to marital status. This gives us something equivalent to a stacked bar chart.

Of course we can do the same thing with ordinary vertical histograms, but with a degree of caution. Since the rectangles have common boundaries the visual impression is often splintered. It is even more difficult to distinguish between the individual groups than with a stacked bar chart. However, the technique is useful for an overview as above.

"Population pyramid type" diagrams

Per cent smokers in different groups by age and sex

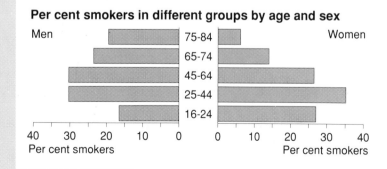

The technique of putting two charts back to back is useful in several contexts. Here we have used it to compare the smoking habits of men and women.

The chart on the left is a bar chart giving the proportions of smokers of different ages. We have calculated the percentage *within* each class, so the fact that the classes are of different sizes does not present a problem. To prevent the chart from being misread we have emphasised both in the title and on the x-axis that it is showing the *proportions* of smokers.

Frequency polygons

Frequency polygons are an alternative to histograms for illustrating a continuous variable. Frequency polygons can be seen as histograms which have been smoothed out and this smoothing out can be an advantage. The histogram with its rectangles can give the impression of dramatic changes at the class boundaries.

Frequency polygons are constructed from an imagined histogram such that the centers of the upper sides of the rectangles are joined together by lines. However, we have problems with the highest and the lowest classes. Even if there are solutions, the problem is so great that we believe the use of frequency polygons should be avoided.

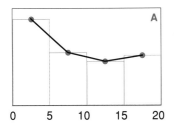

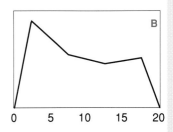

Chart A shows how we construct a frequency polygon for the age data in table B on page 30. It also shows that we do not have any good solutions for the end classes. Taking the line down to the origin makes the area of the lowest class too small, but to give the correct area we would have to take the line down to minus 2.5 years, which is absurd. There are similar problems with the highest class. In spite of everything, chart B is the least misrepresentative.

Comparing several sets of continuous data

If we want to compare two or more categories regarding a continuous variable, such as women and men with regard to age, we cannot use any kind of "grouped histogram". We cannot put the rectangles next to one another without damaging the characteristics of the x-axis.

In a case like this frequency polygons can be a solution. By drawing frequency polygons for the two categories in the same chart we give a direct comparison of ages. Chart C shows this solution for the data in the table on the right. Chart D shows an alternative which is often better, namely placing two histograms one above the other.

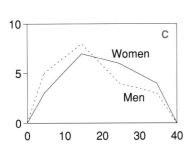

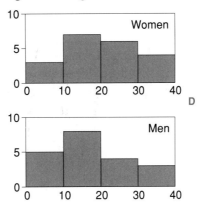

With different sized classes the same applies as for histograms, i.e. that we have to convert the frequencies

Age	Men	Women
0 - 9	5	3
10-19	8	7
20-29	4	6
30-39	3	4
Total	20	20

Stem and leaf charts

Stem and leaf charts are working charts which are reminiscent of horizontal histograms. For small sets of data they offer the possibility of retaining the observation values while also giving the overall picture of a chart. We get a cross between a table and a chart.

Charts C, D and E describe the same set of data for 20 men. By splitting the figures for each observation so that the multiples of ten are transferred to the y-axis and the units are put in the chart we get the stem and leaf chart E. We can now see whether certain values predominate and i.e. that there are no values above 34.

Age distribution for 20 men

E

0	35668
1	03445578
2	1258
3	134

Stem-and-Leaf was introduced by John Tukey

The ages of the twenty men are ranked as follows: 3, 5, 6, 6, 8, 10, 13, 14, 14, 15, 15, 17, 18, 21, 22, 25, 28, 31, 33, 34

The step function graph illustrates the set of data in the table on page 24

Occupants	No.	Cum%
1	1,324,766	36
2	1,150,976	67
3	498,189	81
4	493,408	94
5-	203,001	100

The ogive illustrates the set of data in table B on page 30

Age	No.	Cum%
0 - 4	2,231	37
5 - 9	1,368	60
10 - 14	1,132	79
15 - 19	1,291	100

Although it is difficult to determine the precise size of the sectors with the eye, it is possible to see the relative sizes

Since pie charts do not have a scale axis it is a good idea to give the numerical values in the diagram

Step function graphs and ogives

Until now we have shown "ordinary" frequencies - numbers or percentages. Sometimes we also want to use charts that show cumulative frequencies.

A cumulative frequency states how many observations are *the same as or less* than a given value. Correspondingly a relative cumulative frequency gives the *proportion* of observations which are the same as or less than a given value.

Discrete variables are illustrated using step function graphs. This chart is constructed by adding up all of the frequencies, absolute or relative, up to a certain value. These values are connected by "steps" at the values which the variable can adopt. Continuous variables are illustrated using ogives. Here we add up all the frequencies up to the respective class boundaries. We form coordinates from the class boundaries and frequencies and the points are joined by straight lines. In the ogive below the points (0,0), (5,37), (10,60), (15,79) and (20,100) are joined.

Step function graph
Cumulative percentages

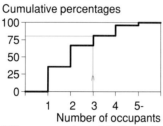

What proportion of the households have up to three members? We see from the diagram that this proportion is about 80%.

Ogive
Cumulative percentages

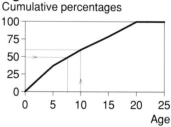

What proportion of the young people are under 10 years old? We see that this proportion is about 60%. We also see that 50% are under about 8 years of age, so the median = 8.

Pie charts

Pie charts are suitable for illustrating percentage distributions of qualitative variables and are an alternative to bar charts. Since it is difficult for the eye to read off precise measurements from a circle, this type of chart is best suited for overviews.

Pie charts are simple to construct, not least with the aid of a computer. Since the total area is 100%, 1 per cent corresponds to 3.6 degrees. The different sectors are placed in the same order as in a bar chart, i.e. retaining the order between the variables. Pie charts begin at "12 o'clock" and are read clockwise.

Consumption of water in Sweden in 1985

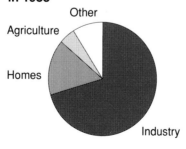

Heating systems in detached houses in 1990

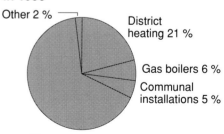

We should not have too many sectors in a pie chart. Five or six is a reasonable upper limit for a lucid chart. As seen in the pie chart above on the right, it is not necessary to use different patterns for the sectors. If we do have different patterns they should go from dark to light. Since pie charts give an overview we often have an "other" class at the end. This should have a low-key pattern so that it does not dominate.

Pie charts for several groups

If we have several groups which we want to compare using pie charts then we draw a pie chart for each group. By making the area of the circle proportional to the size of the group, we can compare both the sizes of the groups and the percentage distributions within the groups at the same time.

When calculating the size of the circles we must remember that the area of a circle is proportional to the square of the radius. Therefore a group which is twice as large is not represented by a circle with a radius of twice the size, but by a circle with a radius about 41% larger. (The square root of 2 is about 1.41.)

Transports within Sweden by transport distance and mode of transport in 1990

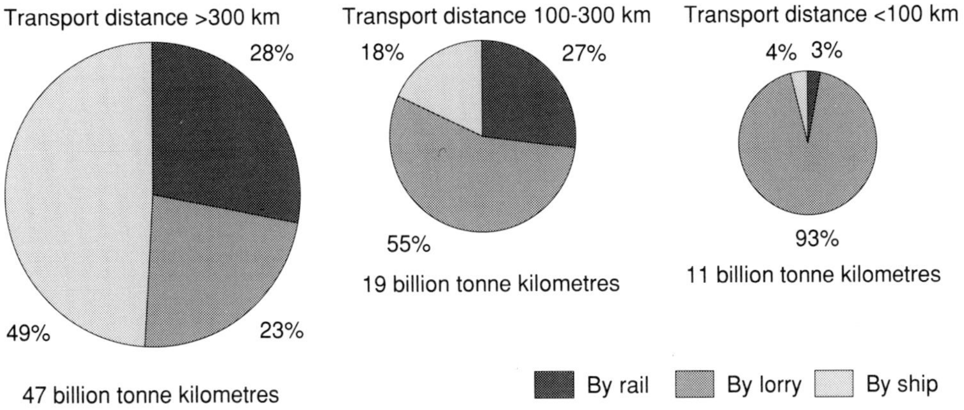
Transport distance >300 km
28%
23%
49%
47 billion tonne kilometres
Transport distance 100-300 km
18%
27%
55%
19 billion tonne kilometres
Transport distance <100 km
4%
3%
93%
11 billion tonne kilometres
By rail
By lorry
By ship

To calculate the radii of the circles on the left we can set the radius for the circle for short-haul transports to 1. The medium-haul transports would then have a radius equal to the square root of 19/11 and the long-haul transports would have a radius equal to the square root of 47/11

If you do not want to show the variation in size, either because it is not interesting, or because the variation is so big that it is practically impossible, then make the circles the same size. Irrespective of whether the circles are made the same size or not, you should give the number on which the percentages are based for each circle.

Variations on pie charts

Most computer programs for statistical graphics have the capacity to explode sectors or to begin the diagram in a different position to "12 o'clock". However, there is never any reason to begin anywhere other than straight up and there is seldom any reason to explode one or more sectors.

If we do explode a sector this should be a signal to the reader that there is something very special about that particular sector. In the diagram below we want to indicate that we have made a special study of half of the non-respondents and either persuaded them to give us information or made a note of the reason why they did not respond. The bar chart on the right of the diagram refers only to the sector which has been exploded.

The rate of response in a specific survey and a follow-up of non-respondents

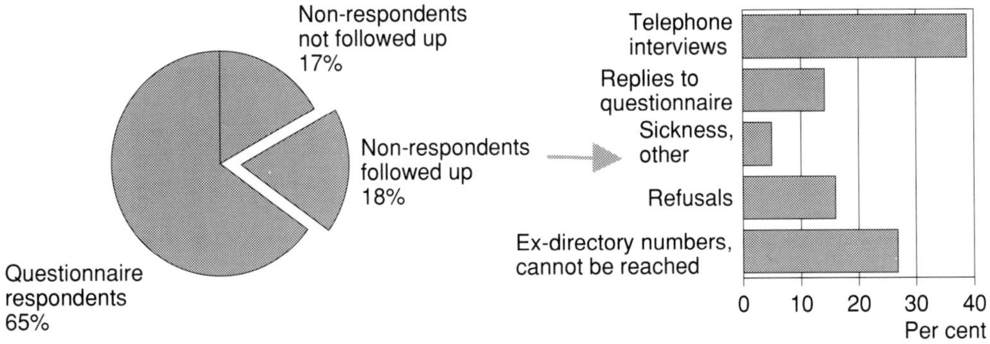
Non-respondents not followed up 17%
Non-respondents followed up 18%
Questionnaire respondents 65%
Telephone interviews
Replies to questionnaire
Sickness, other
Refusals
Ex-directory numbers, cannot be reached
0
10
20
30
40
Per cent

6 Showing development over time

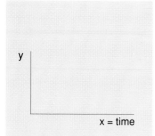

y

x = time

Different time series values are joined by *lines* which form a *curve*

When you want to describe time series in charts it is important to design the chart so that it is easy to make comparisons over time (x-direction). The patterns in time series are generally very difficult to discover by studying tables. When you work with time series you should make several different working charts in order to discover important changes and interesting patterns. Afterwards it will only be possible to convey the patterns resulting from these trials through well-constructed charts.

Bar charts or line charts?

Charts with vertical bars are best suited for vertical comparisons. In a line chart it is easier for the eye to follow the curves for different series and so get a picture of the development over time. Line charts are suitable when you want to judge the *gradients* of curves and are appropriate for the following types of analytical problem:

- In what periods were the changes large?
- When were the turning points?

Bar charts emphasise individual points in time

Bar charts are only suitable if you want to describe a *small number* of points in time

Bar charts are unsuitable if you have several series in the same diagram

Oil supplies to end users 1979-1990

Millions of m^3 per year

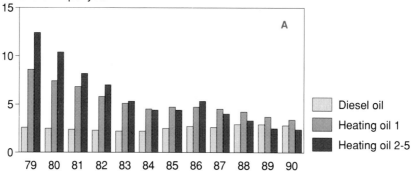

This chart is dominated by the large number of vertical lines and is therefore difficult to read. It is perhaps possible to compare the three types of oil for a single year. However, it is difficult to follow the development over time and see the differences between the patterns. In addition, in a bar chart with several series the time positioning will be inaccurate. The three bars which refer to the same year are displaced along the time axis.

Line charts emphasise development patterns

Line charts are suitable regardless of the number of points in time

Line charts are suitable even when you have several series

Oil supplies to end users 1979-1990

Millions of m^3 per year

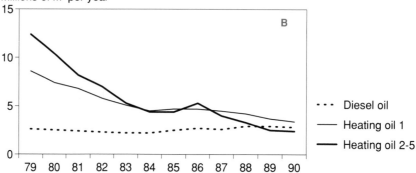

The line chart gives a clear picture of the development – it is easy for the eye to follow each of the three curves over time and to recognise important changes. It can clearly be seen that heating oil 2–5 was the largest product before 1984 and the smallest after 1988, and also that 1986 was a deviating year with a temporary rise.

The building blocks of time series charts

In a line chart with several time series which do not intersect, the various series can have the same line pattern if the legend is written close to the respective curves. If the series intersect each other you have to choose patterns so that the reader easily can follow each series. The legend should be placed so that the picture of the development pattern is not disturbed. If you place the legend in the plot area consider using short names and placing the text where is causes least disturbance, as in chart C. In most of the charts in this chapter we have placed the legend outside the plot area and close to the end of each series. In chart B we have also reproduced the line pattern in the legend because the series lie close.

Horizontal grid lines are generally needed to make it easier to judge gradients and easier to read levels. Vertical grid lines help the reader to determine the timing of changes and turning points. In charts with periodical data (e.g. monthly data) the different years should be marked off by vertical grid lines. So that the grid lines do not obscure the time series curves there should not be too many of them and they should also be thin. The time series curves should dominate and should therefore have distinct patterns.

If you are successful in your choice of patterns for curves and grid lines, it is unnecessary to mark the time series values with symbols. Symbols overload the chart, especially if you have several series which intersect. Chart D is an example of this.

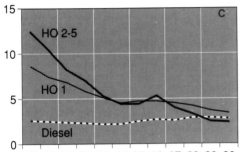
Distinctive curve patterns and discreet grid lines. The legend does not intrude.

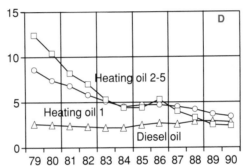
The symbols and legend disturb the time series patterns. The grid lines are too strong.

Time axes

As a rule time series values relate to *periods*, e.g. years, quarters or months. The time axis should be divided in such a way that the ticks separate the periods. The time series values should be placed in the middle of the period they represent. Avoid creating charts with annual data where the ticks and grid lines pass through the middle of each year. The first and last year's values would then end up at the edges of the chart and only take up half the space on the time axis compared with the other years. We recommend that you deal with annual and periodical data in the same way, except when you have series with *many* year values when the grid lines can go through the middle of the years.

Annual data:

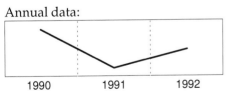

Quarterly data:

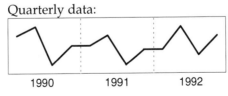

Less suitable for annual data:

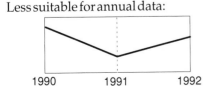

Suitable for annual data with many years:

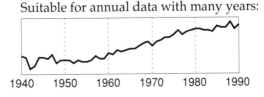

The building blocks of charts are discussed at length in chapter 3. Here we are developing what is most important for time series charts.

The frame should not be allowed to conceal data. There should be space between the data and the frame:

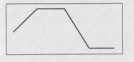

Unsuitable chart:

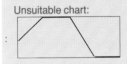

Abbreviations in charts should be explained in an annotation to the diagram

If time series values relate to *points in time* the values should be located with respect to the point in time. If, for example, the values refer to the 31/12 of each year, do as follows:

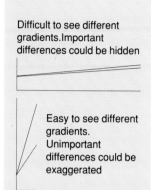

Difficult to see different gradients. Important differences could be hidden

Easy to see different gradients. Unimportant differences could be exaggerated

Proportions of time series charts

The proportions of a time series chart are very important for how the reader perceives the chart's message. The chart should show changes over time. The reader should be able to compare gradients of various curves and see how these gradients change. The gradients are easier to see if the y-axis is *enlarged* and the time axis is *shortened*. It is important not to hide important changes or exaggerate unimportant changes. You should therefore be governed by the purpose of the chart when you decide the proportions.

In the example below we will describe Swedish electricity consumption. Since on average one nuclear power plant corresponds to around 5 TWh per year we want to design the chart so that a change of this order of magnitude can be clearly seen.

The time axis

The chart's time axis is determined by the *choice of time period* and the *length of the axis*. Many time periods per centimetre make it easy to see changes. In chart A the time axis is enlarged compared with B so that each year corresponds to twice the distance. So the gradients which describe changes over time are halved and are more difficult to see.

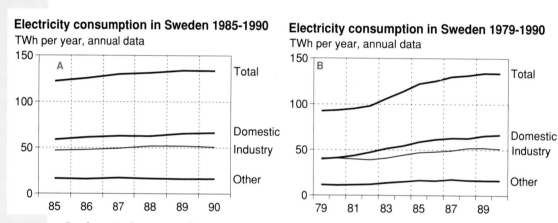

In chart A almost no changes can be seen. The time axis is too drawn out and the y-axis is too squashed – a change of 5 TWh is difficult to see. In chart B it is possible to see certain changes, at least for the whole series. In addition, chart B shows development over a longer period of time. Even if the aim is to show current values, it is important to be able to make historical comparisons.

The y-axis

The diagram's y-axis is determined by the *lowest* and *highest values* on the axis plus the *length of the axis*. Few y-values per centimetre make it easy to see changes. You decide what should be highest and lowest values when you decide what series are to be included and when you decide whether the zero should be included or not.

Compared with A the time axis and the y-axis have been changed so that the gradients have been increased by 7.5 times

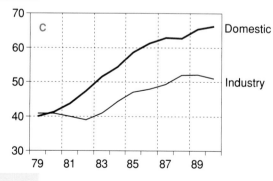

To be able to detect the changes in the three sub-series in chart B the y-axis has to be greatly enlarged. To enlarge the whole chart in the y-direction would give a clumsy result, so we choose to describe the series in separate charts and not to include the zero on the y-axis. The series which were in the middle are shown in chart C. It is now possible to see that the industry series has risen by around 10 TWh and the domestic series by around 25 TWh and also to see *when* the changes occurred.

The example on the previous page illustrates a common situation when one wants to describe series at very different levels. In such situations it is hard to find a y-scale which is suitable both for series with high values and series with low values. The problem can be solved in various ways. One way is to make several charts with different y-axes. Other ways are to use index charts and semilogarithmic charts. We will elucidate this problem and compare different methods in chapter 13.

Should the zero be included on the y-axis?

For time series charts it is important that the reader can make comparisons in the time direction. As we have shown the proportions between the y-axis and the time-axis has great significance for what comparisons are possible. The ideal is to be able to make comparisons in both the vertical and time directions in the same chart. The oil chart fulfils this ideal, the y-axis begins at zero without it being difficult to see changes over time.

The oil chart is chart B on page 36

In other cases it is impossible to make a chart which allows one to make comparisons *both* in the vertical direction and the time direction. You must then decide on priorities and design the chart so that the reader can make the most important comparisons. If the y-axis becomes too compressed when you include the zero, you must enlarge the y-axis so that it does not start at zero. All of the measures which were taken to change chart A to chart C were aimed at making it easier to make comparisons in the time direction.

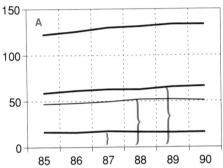

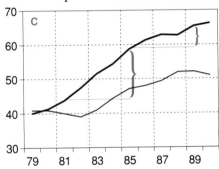

In this chart the reader can only make comparisons in the vertical direction – the diagram only shows differences in levels between the four series, while it is difficult to detect changes over time.

Here we can only compare gradients. The comparisons which are marked with brackets give interesting information – the growth was stronger before 1985 than after. In the lower series cyclic fluctuations can also be seen.

In chart C it is clear that the y-axis starts at 30

If you wish you can bring attention to the fact that the y-axis does not start at zero by cutting the frame and grid lines of the chart in the way shown below:

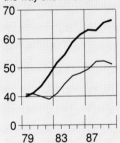

All of the grid lines must be broken. The chart below is unsuitable:

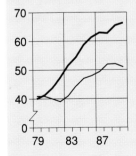

Whether or not the zero on the y-axis in a time series chart should be included is decided by the purpose of the chart and what target group it is aimed at. For an audience which is only to be given a general overview chart B, which shows the relationships between the series, may be sufficient. However, a specialised target group with greater knowledge of the subject should be given a report which includes more comprehensive and more demanding charts. In order to be able to show interesting changes over time it is often appropriate not to include the zero on the y-axis so as to allow a good y-scale. You can also combine several charts. If charts B and C are included in the same report the levels and important changes are both shown.

Different authors of books on graphics have had differing views about the zero on the y-axis. Darrell Huff maintained in the 1950s that the zero should always be included and his extreme example was listened to by many. More modern authors, like Edward Tufte and William Cleveland, have a different view – they let the y-axis start immediately below the lowest y-value in the data set. There is a difference between bar charts and line charts in respect of this question – in a vertical bar chart comparisons are made in the vertical direction and so the y-axis should begin at zero.

If the time series includes values which are zero, let the y-axis start below zero so that no values are touching the frame. The frame of the plot area should not conceal data.

Index charts

Index numbers are comparisons *relative* to a base period. The advantage of this is that one can compare relative changes between series even if the series are at different levels or if they are measured on different scales. The disadvantage is that index series do not give information about differences in levels.

Oil supplies to end users 1979 -1990

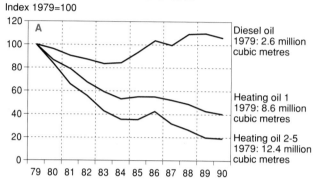

Industry´s electricity consumption 1979 -1990

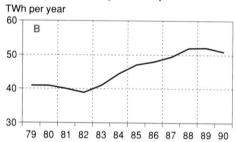

Industry´s production volumes 1979 -1990

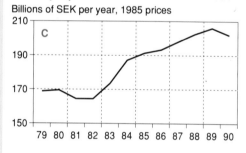

Industry´s production volumes and use of electricity

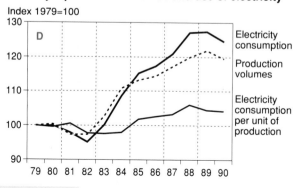

Series at different levels

Readers unfamiliar with statistics could easily misinterpret this as differences in levels between the series. You should therefore add a note giving the levels for the base period. Index charts are harder to interpret than charts on the original scale and are therefore suited to a more qualified target group.

Chart A is an index chart corresponding to the charts on page 36. The scale on the y-axis in the index chart is such that even the variations in the diesel series stand out.

Series measured on different scales

We will compare three series with different scales: industry's electricity consumption (TWh/year), industry's production volumes (billions of SEK/year) and electricity consumption per unit of production (TWh per billion SEK). The two first series are described in charts B and C, the third series, electricity consumption per unit of production, is the ratio between these.

In chart D the three series have been converted to indices and the chart shows that electricity usage rose faster than production volumes after 1984. This is impossible to see in charts B and C where the series are described in their respective original scales.

Sometimes two series are drawn on the same chart using the left-hand y-scale for one series and the right-hand y-scale for the other. Such charts give neither relative nor absolute comparisons. In chart E the reader is given the false impression that production volumes rose fastest.

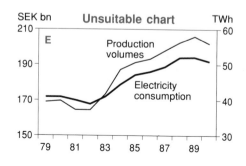

Base year and the index value 100

The index value 100 is the reference level – each value for the respective series is compared with this level. Therefore the index value 100 has the same role as the zero on the y-axis in a usual line chart.

The index should be calculated so that the first period on the time axis is the base period with the index value of 100. In chart F, which is based on the same data as chart D, the first year is not equal to 100, which means that the chart is more difficult to read, the series intersect each other "twice as often" and the values for the base year are difficult to interpret.

Unsuitable base period

Index 1985=100

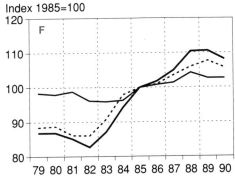

Often you have to convert published indices yourself so that the first period in the chart is 100

The scale value 100 should always be included on the y-axis

Semilogarithmic charts

With semilogarithmic charts the y-axis is in a logarithmic scale and the x-axis is in a normal scale. This type of chart is suitable when you want to study percentage changes. When the y-axis is made logarithmic, the scale steps represent equally large *relative* changes. In the chart below the two different ways of describing the same set of data are compared. They highlight different characteristics in the time series.

Normal time series chart

Fictitious data
Millions of SEK

Semilogarithmic time series chart

Same fictitious data
Millions of SEK

It is often appropriate to work with base 2 logarithms rather than base 10 logarithms

With base 10 logarithms chart H would have scale steps from 10 to 10,000, which would be unwieldy

In chart H each scale step corresponds to a doubling, the scale steps 64, 128, 256, ... correspond to base 2 logarithm steps 6, 7, 8, ...

The base 2 logarithm for y is calculated as follows:
$$^2\log y = {}^{10}\log y / {}^{10}\log 2$$

*This chart illustrates **absolute** changes – of the solid curves, the upper and lower curves have the same and the middle curve a stronger absolute growth in millions of SEK per year. The upper broken curve has larger amplitudes in millions of SEK than the lower one.*

*This chart illustrates **relative** changes – of the solid curves the two lower curves have the same relative growth (10% per year) and the upper curve has lower growth (5% per year). The two broken curves have the same relative amplitudes calculated in percentages.*

Both index charts and semilogarithmic charts illustrate relative changes. Semilogarithmic charts have the advantage that the differences in levels between the series is maintained, but the disadvantage is that a logarithmic scale is difficult to read off for values between the scale ticks. This chart type is suitable if you are addressing a qualified target group which can assimilate the content of the chart.

Each *y* observation in a periodical series can be divided as follows:

Y = trend + economic cycle + season + incidentals

Trends and economic cycles give long-term variations, while seasons and incidentals give short-term variations

In charts with periodical data you should have grid lines between each year

Periodical data

When you want to illustrate a time series with periodical data you must pay attention to the following:

- How are time series with *many* observations illustrated? Eleven years with annual data is eleven observations, but eleven years with monthly data is 132 observations. This means that charts with periodical data can become overburdened with details.
- In general periodical data is *more varied* because it includes both seasonal variations and other short periodical variations. This means partly that it is difficult to find a suitable scale on the y-axis (large variations give many y-units per centimetre) and partly that different variation patterns can conceal one another.

What variation patterns should be shown?

You must decide whether you want to show original, unrefined values which include all variation patterns, or whether you want to show *seasonally adjusted values* which include incidental variations and variations due to trends and economic cycles. It is also possible to show *smoothed values* which are an estimate of only those patterns due to trends and economic cycles. The purpose of the chart and the target group decide what is most important. If there is reason to show both seasonal patterns and long-term development, you should draw two charts.

Chart A below shows a series of unrefined values with all its details. The strong seasonal fluctuation dominates and hides the other patterns. The chart may be suitable if the main purpose is to show seasonal variations.

Industry´s order inflow, volume index 1979-1989
Index 1980=100

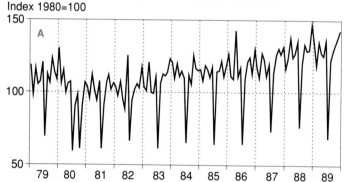

Seasonally adjusted values are often illustrated as in chart B. Since the incidental variations are stronger in this series, the picture gives a restless impression with its 131 lines with different slopes. The chart shows many details, but does not give an overview. The long-term development is difficult to distinguish.

Industry´s order inflow, volume index, seasonally adjusted values
Index 1980=100

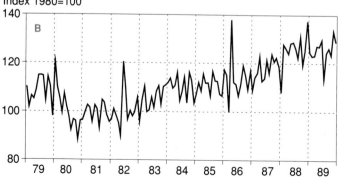

In chart C the long-term development is shown with a curve for the series of smoothed values. In addition the seasonally adjusted values are included as the end points on the vertical lines. The distance between the end points and the curve describes the size of the incidental variation – the variation which is specific to each point in time.

We recommend chart C, which gives both a clearer picture of the long-term development and clearer detailed information about individual points in time than chart B. In addition, it is easier to determine the timing of the observations in chart C because it is easy to follow the vertical lines. Chart B contains 130 angles which give no information and are therefore superfluous. Chart C, on the other hand, contains no graphical elements which do not contain information.

Industry´s order inflow, seasonally adjusted and smoothed values

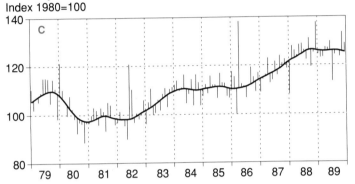

Index 1980=100

We suggest that this type of chart should be called a *needle chart*

The needles which relate to the seasonally adjusted values should be drawn more thinly than the curve which relates to the smoothed values

Needle charts are most suitable when you want to describe *one* time series with periodical data

Several series in the same chart

Chart D is overburdened. It is impossible to show all the details for the respective series and at the same time give a clear picture of the relational pattern. In chart E the details have been filtered out so that only the smoothed values for each series is reproduced. Here the differences between the cycles of the series are clear to see.

Industry´s order inflow and production, seasonally adjusted values

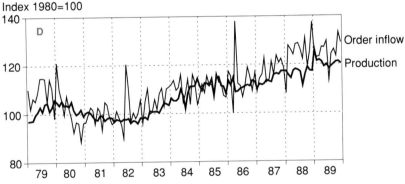

Index 1980=100

Industry´s order inflow and production, smoothed values

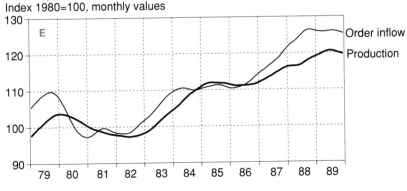

Index 1980=100, monthly values

Adjusting and smoothing values makes the y-scale less cramped. In chart A, which has non-adjusted values, the y-scale covers 100 units, but in chart E, which has smoothed values, the y-scale is only 40 units long

Accumulated line charts – area charts

Compare with the stacked bar charts in chapter 4

In many cases it is necessary to describe time series which consist of a sum and its parts. If there are only few parts and they have simple development patterns, it may be appropriate to draw an accumulated line chart, which illustrates the whole and its division into parts.

How does an area chart work?

Chart A describes the development of employment as a total and divided into two parts, the categories men and women. The total is described by the top curve and the parts by the areas with different patterns. The number of employed men has remained almost constant over the time period and because this series is placed at the bottom it is easy to see that the employment of women has gradually increased.

In accumulated line charts it is always easy to read off the total and the lower part. However, it is difficult to read the other parts if the lower part varies greatly. Chart B is an example of this; it is almost impossible to see how the number of unemployed women has varied over time.

Differentiate between area charts and line charts

You should design your area and line charts so that your readers do not confuse the chart types. In area charts you should use patterns or shading, but not in line charts. The data labels should be linked to curves in line charts and to areas in area charts. If there is a risk of misunderstanding, you can use brackets as in chart A. It is often appropriate to mark the curve for the total with a legend.

We found the equivalent of chart C in a daily newspaper. Chart C gives the impression of being an area chart, but from the text we understood that it was a line chart where the upper *curve* corresponded to the percentage of unemployed and the lower *curve* corresponded to the percentage of people in job creation schemes. Those who designed the chart were probably thinking three dimensionally with the curves as two slices where the slice for the unemployed was at the back. The result was a chart which was probably misunderstood by many readers.

Number of employed people 1970-1990
Millions

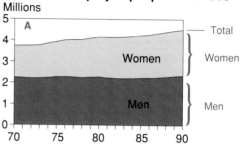

Number of unemployed people 1970-1990
Thousands

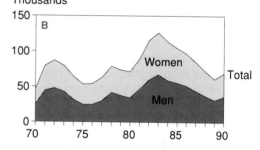

Unsuitable chart

People unemployed and in job creation schemes

Per cent

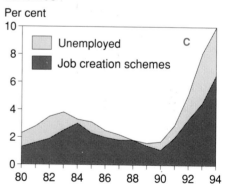

The chart includes actual values up to and including 1992 and forecasts for 1993–94. For 1994 a total of 16.5% are expected to be outside the regular job market and 6.5% to be employed through job creation schemes.

Area charts or line charts?

In more complicated situations with many parts or with several parts which change at the same time, you should avoid area charts and use ordinary line charts instead. Chart D is an example of a difficult-to-understand area chart. The drop ringed by the blue circle could be misinterpreted as a decline for the commodity groups 7–9, even though only group 3 has fallen. This is shown clearly in the ordinary line chart E, which shows only the parts and not the total.

Imports divided by commodity group

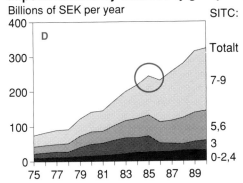

Imports of various commodity groups

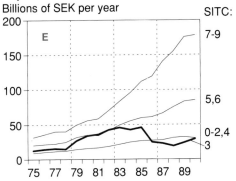

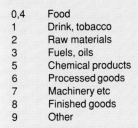

Commodity groups within SITC:

0,4	Food
1	Drink, tobacco
2	Raw materials
3	Fuels, oils
5	Chemical products
6	Processed goods
7	Machinery etc
8	Finished goods
9	Other

Area charts for proportion

Accumulated line charts can also be used to show how *proportions* change over time. This type of chart demands great care when deciding the order in which the parts are to be placed. Place the parts which have the smallest changes at the bottom. In chart F we have placed groups 0–2 and 4–6 at the bottom. In the period 1979–85, when the oil prices were high, group 3 increased its proportion of the import value at the expense of groups 7–9. However, an ordinary line chart like G is easier to construct and to interpret, even when dealing with proportions.

Proportional charts can be difficult to interpret unless they are combined with charts on ordinary scales such as charts D or E above. When you illustrate proportions, information about the development of the total is lost.

Imports divided by commodity group

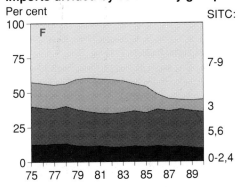

Imports of various commodity groups

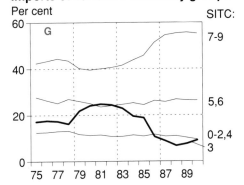

7 Showing relationship

Scatterplots

Scatterplots are used to show relationship (causal relationship or covariance) between two *quantitative* variables. The data consists of a number of pairs of coordinates (*x, y*). Each pair of coordinates is indicated by a dot or circle in a system of coordinates.

Chart A shows the relationship between income and the consumption of food. The data consists of income (*x*) and food as a proportion of total consumption (*y*) for 100 households. The household indicated by blue lines has income of 73 thousand SEK and the proportion of food is 9.5%.

It is often useful to include an estimated regression function as a curve in the chart in order to give an overall picture of the relationship – the dots show the details.

Aggregated data

There are two cases you should distinguish between even if the charts have similar appearances. One is when you have data on the *object level* and the other is when you have *aggregated data* for groups of objects. Chart B shows the corresponding relationships for groups of households. Each dot represents the mean values of x and y for a *group* of households.

If you have a large set of data it often becomes too crowded to draw markings for all of the observations. It may then be appropriate to use aggregated data in the chart. When you are working with official statistics using aggregated data is generally the only possible alternative because individual values are not published.

Quantitative   Quantitative

x ➤ y

Income   Consumption
of food

SEK = Swedish crowns

There should be space between the data and the frame

Unsuitable chart

Since the objective is to show relationships and not levels the axes do not need to begin at zero

Food´s part of consumption for households with different incomes
Per cent

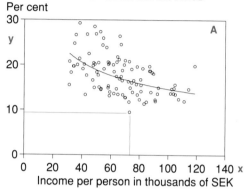

Food´s part of consumption for groups of households with different incomes
Per cent

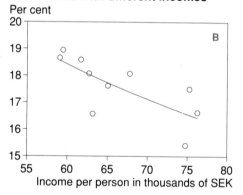

The proportions of the axes and the risk of optical illusions

The scatterplots C and D show the same set of data. Because the dots in chart D are so close to each other, the reader is misled into perceiving the relationships in D as stronger. You should let the data fill the whole of the plot area as in C, i.e. each axis should begin immediately below the lowest value and end immediately above the highest. The axes are then in proportion to the variation in the variables and the relationship is clearly visible, while the risk of optical illusions is small.

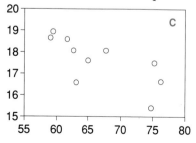

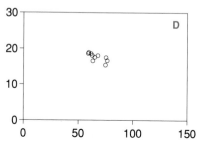

Several x-variables

The simplest scatterplot describes the relationship between a quantitative x-variable and y. By using symbols or by drawing several charts you can show the relationships between several x-variables and y. Since you should avoid having too many symbols or charts the corresponding x-variables should have a limited number of classes.

In charts E and F the proportions of unemployed are shown for men and women of different ages. The variable "age" is the quantitative x-variable and should therefore form the x-axis of the chart. The qualitative variable "sex" is included by the use of two different symbols in the chart. In order to emphasise the relationship between age and unemployment you can join the dots as in chart F. The scatterplot then changes into a line chart.

Per cent unemployed by age and sex 1991

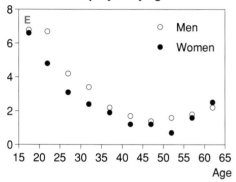

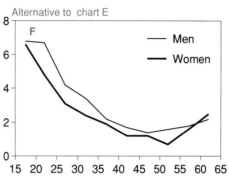

Alternative to chart E

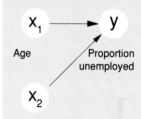

A qualitative variable with two classes can be treated like a quantitative variable with the values 0 and 1. The variable 'unemployment' is treated in this way here

Quantitative   Quantitative

x_1 ⟶ y

Age   Proportion unemployed

x_2

Sex (qualitative)

Charts of the same type as chart E become difficult to interpret in two situations:
- Many classes for the qualitative x-variable result in too many different symbols which can be difficult to distinguish between.
- With a large set of data the different clusters of dots could overlap to a large extent. It then becomes difficult to distinguish between the different categories.

In chart G below the objective is to show how the proportion of consumption accounted for by food varies with the income and size of the household. Four symbols distinguish between households with 2, 3, 4 or 5 members. Because the chart is difficult to read it is useful to draw the four small charts instead. In order to be able to compare these charts they should be placed close together and have identical scales. The charts should also have common reference lines describing the relationships for the various subgroups.

Instead of depicting all data in one complicated chart it is often advisable to make a system of small charts where each depicts one part or one aspect of the data

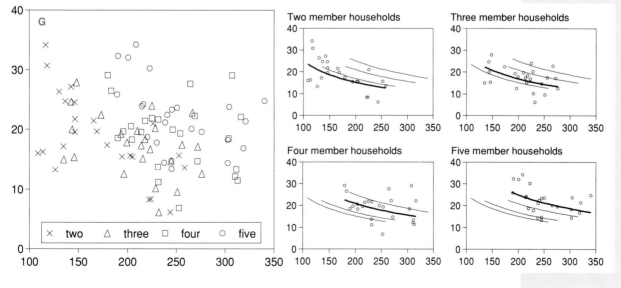

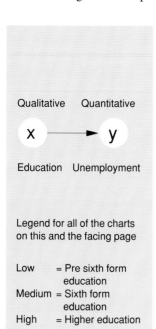

Qualitative   Quantitative

x ⟶ y

Education   Unemployment

Legend for all of the charts
on this and the facing page

Low   = Pre sixth form
          education
Medium = Sixth form
          education
High   = Higher education

Showing relationship with bar charts

If all of the x-variables are *qualitative* the relationship is illustrated using a bar chart.

Chart A shows the relationship between a qualitative x-variable and a quantitative y-variable. The proportion of unemployed (mean value for y) is lower for those with a high level of education.

The grouped bar charts B and C show the relationship between two x-variables and a *quantitative* y-variable. At first the eye compares the bars within the same group. Therefore chart B first shows the relationship between gender and unemployment, while chart C first shows the strong relationship between education and unemployment.

Percentage unemployed for different educational levels 1991

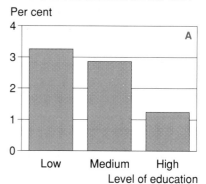

Unemployed by sex and education 1991

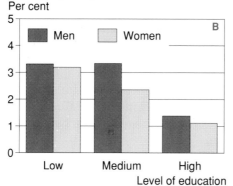

Unemployed by education and sex 1991

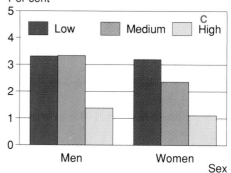

In chart B we have given priority to showing the relationship between sex and unemployment:

In C we have given priority to the relationship between education and unemployment:

When both the x and y-variables are *qualitative* you can choose between a bar chart with stacked or grouped bars. If the x-variable has many classes and the y-variable has few, then a stacked chart like chart D is preferable. If, on the other hand, x has few classes and y many, then you should choose a chart with grouped bars like E. Note that in chart D it is difficult to see the difference between the genders for the middle class. In chart E the reader can easily see this difference.

Qualitative   Qualitative

x ⟶ y

Gender   Education

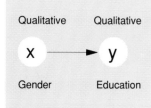

Men and women by level of education 1991

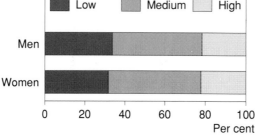

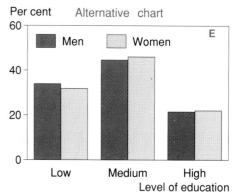

Showing relationship with dot charts

The relational charts we have shown up to now have limited possibilities:
- One or at most two variables
- Few classes, preferably not more than three, for the x-variables

In many situations this is too limited. Multi-way tables with many variables are difficult to read and the patterns of relationships which may exist can be difficult to make out. Dot charts (mentioned in Chapter 4) are a suitable type of chart for these situations.

The basic form of the dot chart describes the relationship between an x-variable and the dependent y-variable. The horizontal scale refers to mean of y or the proportion of y in a certain category. In chart F the x-variable is age and the y-variable is unemployment.

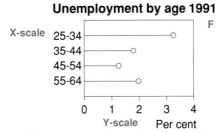

Unemployment by age 1991

In charts G-J we have added the variable "sex" in different ways. Chart G consists of two diagrams, one for each sex. The relationship between sex and unemployment can be clearly seen but it is more difficult to see the relationship between sex and unemployment since the reader then has to make comparisons on two y-scales. In chart H we use symbols for the sex variable, while in chart J we have grouped the data in order to be able to introduce the sex variable. In these two charts it is easy to see how both x-variables co-vary with y since all of the comparisons are made on the same y-scale.

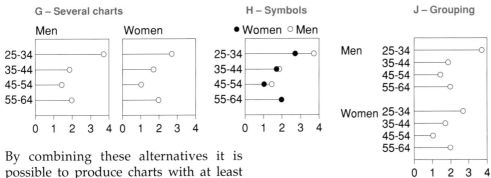

By combining these alternatives it is possible to produce charts with at least four x-variables. In the chart on the right we have made it a priority that the difference between 1990 and 1991 should be clearly visible. The age differences and differences in levels of education can also be clearly seen. In order to facilitate comparisons between male and female unemployment we have added grids lines.

If you have x-variables with many classes then you should group them as we have done with age and education. The chart illustrates the following relationships:

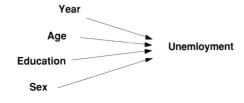

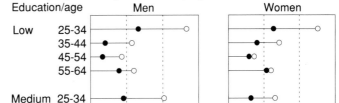

Unemployment by education, age and sex 1990-1991

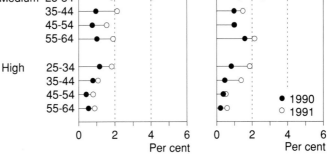

Our method of using dot charts to show relationships was inspired by William Cleveland's book *The Elements of Graphing Data*

Different types of chart make different demands on the eye's ability to perceive differences. We return to this question in chapter 11

We suggest that this type of chart be called a *barometer chart*

Chart A och B illustrate:

Age ——→ **Unemployment**

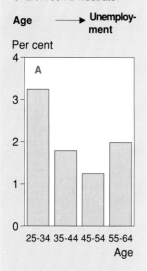

Chart C och D illustrate:

Sex

Age ——→ **Unemployment**

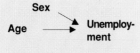

Chart E och F illustrate:

Time (years)

Sex ——→ **Unemployment**

Age

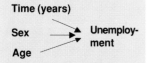

Showing relationship with barometer charts

Earlier in this chapter we have used bar charts or dot charts to describe relationship. For example, the relationship between age and the proportion of unemployed is described by bar chart A in the margin. If we remove the axis the bars are standing on and only indicate the heights of the bars with dots then we get barometer chart B.

Barometer charts can also be used to show y-differences for x-variables with *two* classes. The lines in chart C emphasise the differences in unemployment between men and women and it is clear for which age class the differences are largest or smallest. Chart D is a variation of chart C where the scale values on the y-axis have been replaced by observed y-values. Chart D is therefore a combination of table and chart.

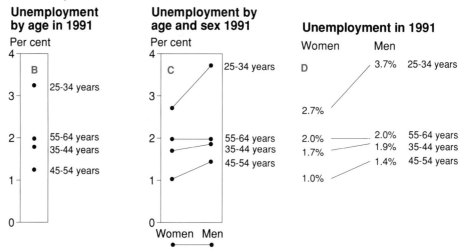

By using different symbols we can insert an additional x-variable. The barometer chart E shows changes in unemployment for eight groups. Unfortunately there are often several intersecting lines in such cases and it can be difficult for the reader to connect the lines to the right groups. In a situation like this, instead of producing a barometer chart which is difficult to read, you should produce a dot chart like chart F.

In chart E the differences are striking but it is difficult to connect the differences to the groups. In chart F the differences are not so obvious but it is easy to see which groups are being referred to.

8 Showing variation

Sometimes you need charts to show variation in a variable or to compare variation between different sub-groups. We have also given examples of chart types which show variation in earlier chapters.

- Bar charts show the distribution of a *qualitative* variable.
- Histograms show the distribution of a *quantitative* variable.
- Time series charts show variation of different time series components.
- Scatterplots show variation of two variables.

In this section we discuss *boxplots* and *Lorenz charts*, mainly as alternatives to the histograms in chapter 5.

Boxplots

Boxplots show both average measurements in the form of medians and variation measurements in the form of ranges and interquartile ranges. The three quartiles and any maximum and minimum values are marked in the chart.

The quartiles divide the data into four groups with an approximately equal number of values in each group. Around 25% of the values are lower than the *lower quartile*, around 50% are lower than the *median* and around 75% are lower than the *upper quartile*. The interquartile range is the difference between the upper and the lower quartile, and the variation range is the difference between the maximum and minimum values.

The various components of the boxplot:

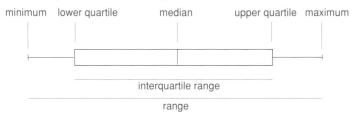

Boxplots can be drawn in different ways, as shown here using the data in the margin:

Boxplot with whiskers

Boxplot with whiskers and extreme values

Boxplot without whiskers

When you want to illustrate a small set of data where all the values are of interest, you can draw a detailed boxplot with both whiskers and extreme values. If you do not want to draw attention to the extremes, you should choose the alternative with only whiskers. The highest value in the example above may be considered to be extremely high.

For a large set of data it is usually best to draw a boxplot without whiskers because the extreme values are probably of little relevance and are not of interest to the reader. If you are working with official statistics regarding individuals or companies, the boxplot without whiskers is the only alternative because individual values are not published.

Boxplots can also be drawn with vertical bars, which may be practical if you are comparing many groups. There are several ways of drawing the shapes of the boxes and the appearance of the whiskers, but here we are only showing the basic types. You will notice that different computer programs calculate the quartiles and judge the extreme values in somewhat different ways. This is only of significance for small sets of data.

A set of data with 11 observations is ranked as follows:
2.0 2.7 3.0 3.1 3.8 4.5 4.7 5.1 5.5 6.0 9.8

The following is therefore true:
Minimum value = 2.0
Lower quartile = 3.0
Median = 4.5
Upper quartile = 5.5
Maximum value = 9.8

There are rules on how to define extreme values (outliers). We refer the reader to John Tukey's book *Exploratory Data Analysis*

Note that quartiles are *not* affected by extreme values

The same data has been used in charts A–C

Boxplot or histogram?

If you want to make a detailed depiction for a quantitative variable (e.g. wages) you should draw a histogram as in chapter 5. But if you want to compare different sub-groups (e.g. men's and women's wages) it becomes difficult to get a clear picture if you draw several histograms because these are so detailed.

In this situation one usually goes to the other extreme and simplifies as much as possible by only describing the mean values using bar charts. The disadvantage with such bar charts is that all the information about the variation of the data is missing.

Boxplots are a suitable type of diagram in such situations, showing variation without being difficult to understand.

Chart A shows that there are differences between average wages and wage spreads, but it is difficult to see the size of the differences. Chart B only shows the differences in average wages, but chart C shows both the average wage difference and that the spread is greater among the men.

Wage distribution within company X 1990

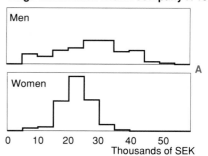

Average monthly wages in company X 1990

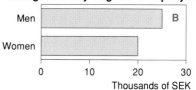

Monthly wages in company X, 1990 quartiles

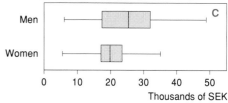

SEK = Swedish crowns

Boxplots for showing relationship

When you want to illustrate relationship you should draw charts which allow many comparisons, and here the advantages of the boxplot become obvious. The example below shows how wages are affected by three factors. You would need 16 histograms for the same description. We have drawn boxplots without whiskers because minimum and maximum values have not been published. Even if they were available, the chart would be overburdened with 32 extra lines which would emphasise eccentric values in the data. By making the wage scale logarithmic, chart E shows *relative* wage spreads, i.e. the spread in relation to level. The measurement of relative wage spreads is of interest for the analysis of wage statistics, but demands a lot of the reader.

Monthly wages for industrial employees 1980 and 1990

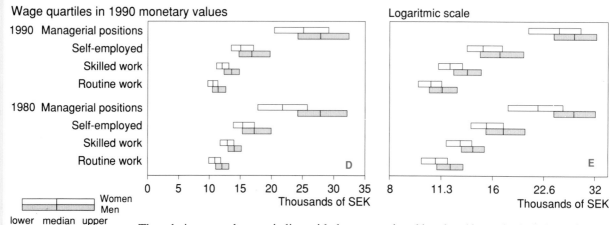

The relative spread grows in line with the occupational level and is particularly large for women in managerial positions. Relative wages have been very stable between 1980 and 1990 with one exception – women in managerial positions have got closer to the men's wage levels.

Lorenz charts

This chart type is based on a method for analysing variation whereby the observations are ranked and a comparison is made of two cumulative distributions. We will explain this using an example.

Ranking from largest to smallest

A company sold goods worth 489 millions of SEK to a total of 253 customers in 1989. These customers are ranked in order of sales value with the largest customers first:

Customer	Sales million SEK	Accumulated shares: Share of customers	Share of sales
A	7.1	1/253	7.1/489
B	6.5	2/253	(7.1+6.5)/489
C	4.8	3/253	(7.1+6.5+4.8)/489

For each customer the accumulated share of customers is put on the x-axis and the accumulated share of sales on the y-axis.

If there was no variation in the data on sales values all customers would be equally large and the points should then follow the diagonal in the chart. The larger the deviation from the diagonal, the larger the variation.

When the same is done for 1992 the variation for the two years can be compared in the same chart.

Ranking from smallest to largest

In the above example customers were ranked from the largest to the smallest. If instead we rank the customers from the smallest to the largest the Lorenz curve will lie *below* the diagonal. This is the usual way to arrange data for the analysis of income distributions.

The chart on the right compares the income distribution in Sweden in 1921 with the corresponding distribution in 1990. The income receivers have been ranked from the lowest to the highest assessed annual income for each year.

Lorenz charts are well suited to economic variables because the conversion to percentages means that differences in monetary value or currencies do not disturb the comparisons.

The large customers' share of sales

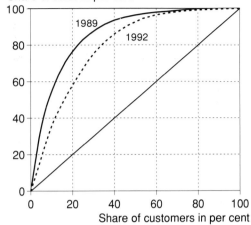

In 1989 20% of the largest customers accounted for around 75% of the company's sales. In 1992 this share had fallen to around 55%. Therefore the variation fell between these two years, which means that the company became less reliant on a few large customers.

Income distribution in 1921 and 1990
(Income receivers with the lowest income and their share of the total income sum)

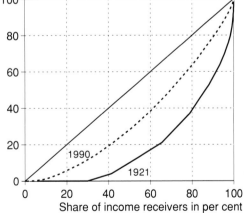

The lowest 40% of income receivers had only 4% of the total income sum in 1921, but around 20% in 1990. The chart shows that variations in income were greatly reduced between 1921 and 1990.

9 Showing flows

By *flow charts* we mean a schematic picture of a process. For example you can give a general picture of production in an industry with a flow chart. If the main aim is to provide statistical information we also use the term *flow chart*. The chart should then give numeric information about the size of stocks and flows between stocks.

You must not include too many flows in the same chart because the chart would become difficult to read with a confusion of intersecting flows. You therefore have to confine yourself to showing the most important flows and exclude or combine the others.

Throughout this chapter we have made the charts as simple as possible by only illustrating flows with simple arrows. We thus avoid the very laborious alternative of arrows whose width is proportional to the flow. Unlike most other chart types we prefer to give figures within the flow chart itself if there is enough space. Thus the chart actually becomes a well-structured table and a graphical picture of relationships between sizes.

Inflows and outflows only

In many cases it is not possible to link a particular inflow with a particular outflow, e.g. in diagrams which show income and expenditure. In other cases there is no information about the links between various inflows and outflows. You can then draw the simplest kind of flow chart without intersecting arrows.

Different units of measurement

If the flows are measured in different units you cannot describe the size of the flow in any way other than by giving figures. We have therefore made all the arrows and rectangles in chart A of equal size.

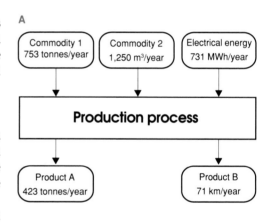

Same unit of measurement

Chart B shows the size of each flow with shaded rectangles where the width of each rectangle is proportional to the respective flow. The total flow is indicated by the chart's outside lines.

Flow charts can also be drawn horizontally as in chart C. This gives you room for more flows and avoids abbreviations in the legends.

Charts of this type are suitable for describing cashflows, i.e. income and expenditure flows. Here we show the flows which describe supply and usage in the US economy in 1992

Expenditure on the GDP in the USA, 1992. Billions of dollars in current prices

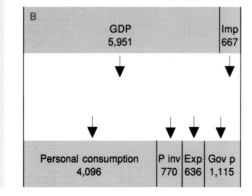

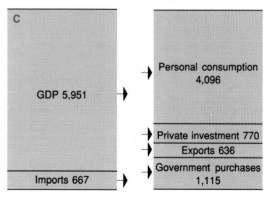

Complete input-output matrices

When you have information which tells how each inflow is linked to the various outflows the statistical data can be combined in a two-way table. The example below is based on information about Sweden's energy balance for 1991.

The energy balance is divided up *simultaneously* into 16 types of energy and eight final usage areas. A complete flow chart should therefore contain $16 \cdot 8 = 128$ arrows or sub-flows (because some flows are equal to zero, the balance sheet contains "only" 71 sub-flows). A chart with that many arrows would be both very laborious to draw and very difficult to read for most users.

We have therefore chosen to limit ourselves by grouping the energy types into three large groups and distinguishing between four usage areas. The table below shows the energy balance after this grouping.

Energy balance, 1991. All types of energy converted to petajoules

Type of energy:	Final use: Transports	Industry	Domestic	Other users	Total
Oil	284	68	84	67	503
Electricity	9	184	147	99	438
Other types	0	242	116	51	409
Total	292	494	346	217	1,350

If we only knew the margins in this table we would draw a flow chart of the same type as charts B and C. But because we have the complete flow matrix we can draw a more detailed chart where we link each inflow (energy type) to its various outflows (final usage).

Chart D, where each sub-flow has been drawn as a "pipeline link" between the energy type and usage, is very laborious to draw. The size of the flow is represented by the width of the pipeline. In chart E the pipelines have been replaced by simple arrows which are easy to draw. The sizes of the sub-flows are shown in chart E by the height of the rectangles. Chart E includes all of the information in the table above. Since chart E is no more difficult to read than chart D we recommend that you illustrate flow matrices with charts of type E.

In chart E we have marked the four largest flows with thicker arrows and the smallest flow with a broken arrow.

Sweden´s energy balance 1991. Energy types by final usage, in petajoules

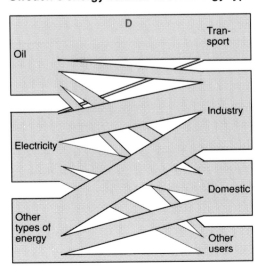

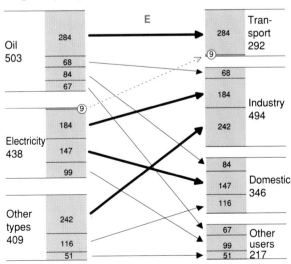

Flows between stocks

When you also have information regarding stocks at two points in time you can draw a chart which shows both stocks and flows. The example below is based on Statistics Sweden's opinion poll in November 1995. In January 1995 Sweden joined the European Union after a decision based on a referendum in November 1994. One year after this referendum the opinion poll was done and the respondents were asked how they voted in the referendum and about their present opinion. At each point in time there are three stocks: Positive to membership ("Yes"), negative to membership ("No") and did not vote/don't know ("?"). Between these two points in time there are also six different flows due to changes in different persons opinions.

The chart below shows stocks and flows as the areas of circles and arrows. The size of the circles is determined in the way described on page 35, i.e. the radii are proportional to the square roots of the quantities to be illustrated. This square root transformation makes it possible to show a very small flow (0.2) in the same chart as a large stock (40.4). The disadvantage is that the stocks at the two points in time are not shown directly. Instead the reader must add (25.4+0.2+1.2 for "Yes") to get the stocks in November 1995. It is a good idea to combine this kind of chart with a table.

Change in opinions regarding membership in the European Union between November 1994 and November 1995
Percentage of those who responded both times

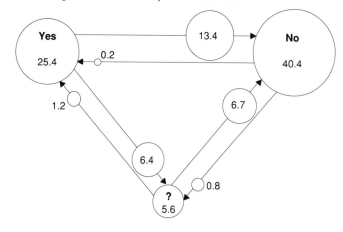

Opinions of membership among the same individuals, November 1994 and 1995
Per cent

Opinion poll	Referendum, November 1994			
November 1995	Yes	"?"	No	Total
Yes	25.4	1.2	0.2	26.8
"?"	6.4	5.6	0.8	12.8
No	13.4	6.7	40.4	60.5
Total	45.2	13.4	41.4	100.0

Geographic flows

Flows between geographic areas may be illustrated by drawing flows on an underlying map. Here again it is best to show the flows with simple arrows.

The chart on the right shows two different ways of depicting geographic flows. Naturally you should only use one of these in the same chart.

The fallout over Sweden is shown with a stacked column where the reader can see the contributions of various countries.

The exchange between the UK and France is illustrated here by marking each flow with an arrow and the size of the flow with circles of different area. Numerical information is also given in the chart.

We have chosen to depict the fallout over Sweden with one column because it is easier to compare rectangles with the same base than different sized circles

Sulphur fallout in Sweden and exchange between the UK and France 1991
Thousands of tonnes

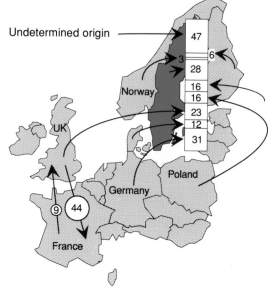

10 Showing geographical variation

Maps have very different appearances and are often grouped into *topographical* and *thematic* varieties. A topographical map depicts first of all that which is immediately visible in the terrain. It shows buildings, roads, forests, etc. together with administrative boundaries. The reader can identify his surroundings from a topographical map.

A thematic map can be seen as a superstructure over a topographical map. The map's principal message gives particular information within a limited subject area. There are, amongst others, geological maps and vegetation maps superimposed on topographical base maps. Other thematic maps, such as *statistical maps*, can be considerably more schematised often without any more geographical features than a simple land outline and some administrative boundaries.

There are many cartographical methods for showing statistical information on different types of statistical maps. This chapter will discuss choropleth maps, square maps, isopleth maps, density maps and cartograms

Statistical maps

Statistical maps are essential where the object is to illustrate location-specific information within an area and are used to show variation, differences and similarities between areas. The maps have administrative divisions such as counties and districts, but they may also have other geographical divisions such as drainage areas and climatic zones. The statistical information is put on the real locations on the map and shown using symbols such as shading, patterns, small charts, etc.

Level of employment 16-64 years, 1991

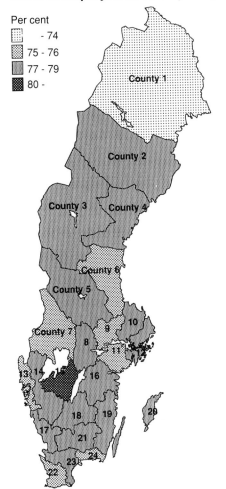

Per cent
- - 74
- 75 - 76
- 77 - 79
- 80 -

The choropleth map, bar chart and table on this page all show the levels of employment in Sweden.

In the table in the margin it is difficult to see any pattern despite the fact that it is sorted on a declining scale from county 15 to county 1. The bar chart is scarcely any better for showing differences.

The choropleth map is the best solution. It is easy to see which county has the highest or lowest level of employment and which counties show similar tendencies.

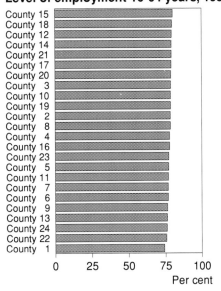

Level of employment 16-64 years, 1991

Level of employment 16-64 years, 1991

County 15	80.11
County 18	79.66
County 12	79.30
County 14	79.24
County 21	79.19
County 17	79.14
County 20	78.83
County 3	78.67
County 10	78.54
County 19	78.53
County 2	78.45
County 8	78.42
County 4	77.90
County 16	77.83
County 23	77.24
County 5	77.16
County 11	76.98
County 7	76.90
County 6	76.83
County 9	76.38
County 13	76.15
County 24	76.09
County 22	75.66
County 1	74.19

Dividing into classes and choosing patterns are discussed on page 61

You should be clear about the fact that statistical maps contain subjective elements to an even greater degree than other diagrams. For this reason the production of maps will never become a mechanical procedure

Unsuitable map
Persons employed 1991 16-64 years

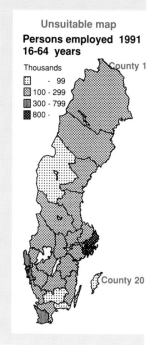

Choropleth maps – showing ratios

Choropleth maps are also called *hatch maps* or *shaded maps*. This type of map is suited to showing ratios such as proportions, intensities and averages for areas.

The ratio which is to be illustrated is divided into classes and is represented on the area's surface by colour, shading, or patterns. The choice of class boundaries and the setting of class patterns are decisive for what information you convey. It is important for you to be clear right from the start about what the map should show.

Choice of administrative divisions

Which administrative division you choose to use (county, district etc.) is of great significance. The district map A on this page and the county map on the previous page both show levels of employment. If you compare the two maps you notice how much more detailed the information provided by the district map is.

You get a greater distribution of ratios when they are shown at the lower level (district) compared with the higher level (county). This means that the class divisions should be different for the county map on the previous page and the district map on facing pages.

On the county map on page 57 county 1 looks as if the whole of the county lies below 75%. When we go down to the district level in map A it is clear that the variation within county 1 is great. A map which is too detailed, on the other hand, gives a poorer overview.

Problems with choropleth maps

One of the characteristics of choropleth maps is that large areas can be over-represented. In Sweden this effect must be born in mind for mountain districts and sparsely populated areas. The large areas could dominate visually over the small ones, despite the fact that the latter may be more important.

Not for showing numbers

Choropleth maps are clearly unsuitable for showing numbers. The map in the margin gives the impression that employment is more sparse in county 20 than in county 1. This is wrong since county 20 has few employed people within a small area, while county 1 in the north part of Sweden has more employed people within a much bigger area.

When we show areas we give the reader the impression that the varying shade of the areas corresponds to varying intensity of the statistical measurement. Therefore, choropleth maps should only be used to illustrate ratios. For absolute numbers you should draw density maps instead, see page 62.

Level of employment 16-64 years 1991

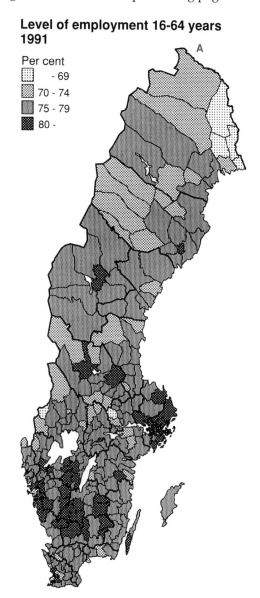

Choropleth maps without borders

In map A on the previous page both district and county boundaries are drawn in. If in a choropleth map we take away the administrative boundaries, the characteristics of the ratios being illustrated can be seen more clearly.

In map B the variations in the level of employment are considerably clearer than if the boundaries had been left in. The disadvantage is that it becomes more difficult to tell where you are on the map when there are no boundaries. A compromise between the two is to draw only the county boundaries on a district map.

This type of map should not be confused with an isopleth map (see page 60).

Detailed areas

If a map contains many small areas it could be difficult to show the statistical information in these.

This is often the case with metropolitan areas. It may then be appropriate to remove these districts from the map and enlarge them, as in map C. On the main map the districts concerned may either be left white or show the relevant patterns.

A problem often arises with detailed areas. How much should be shown? How small can an area be and remain visible? There is no clear-cut answer to these questions except that you should experiment until you find a good balance.

Omitting or minimising the boundaries in a choropleth map is an application of the principle of "letting the data dominate" (see also pages 66–67)

Level of employment 16-64 years, 1991

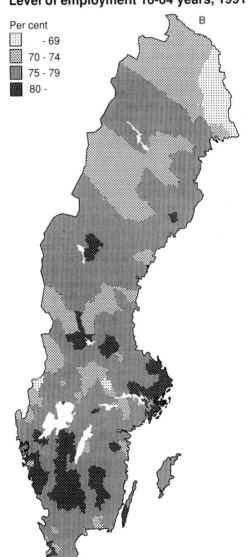

Per cent
- 69
70 - 74
75 - 79
80 -

B

Level of employment 16-64 years, 1991

Per cent
- 69
70 - 74
75 - 79
80 -

C

Metropolitan area 2

Metropolitan area 1

Metropolitan area 3

Square maps

An alternative to administrative boundaries is to divide the map into geometrical areas such as squares and hexagons. The areas are then filled in with the relevant patterns, symbols or numerical data.

Map A is divided into squares of 50 by 50 kilometres. It shows how the pH-value of the soil varies between different parts of the country. The map is based on analyses of 2,200 samples. The most acidic areas are in south-western Sweden.

The size you choose to make the geometrical areas depends upon how close together the measurement values lie and what level of detail you wish the map to show.

Acidification in the soil 1991

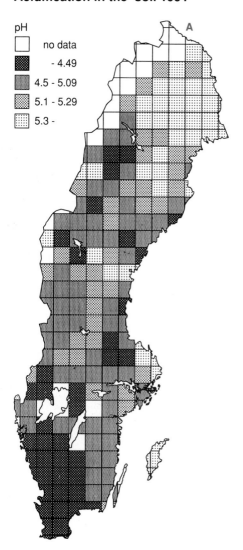

As a rule we use light patterns to show low values of a variable. For the pH values we then get map B below.
Since low pH values mean acid soil we prefer the order in map A where the acidic areas are emphasised

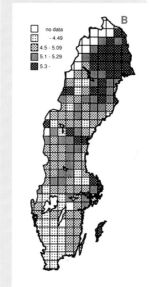

Isopleth maps

For an isopleth map we switch from administrative boundaries or grids to let the characteristics of the statistical variable determine the boundaries directly.

A characteristic which is spread continuously over the whole of the geographical area, for example depth of snow, can be shown by lines which connect points with the same variable values, so-called isolines. Elevation contours (5 or 10 metres) on ordinary topographical maps are examples of isolines. Common isopleth maps are maps of atmospheric pressure and temperature. Creating the boundaries (the isolines) on isopleth maps by processing the measurements is difficult and requires special computer programs.

Map C shows the amount of cesium-137 per square metre according to measurements after the accident at Chernobyl. If we instead chose to show the measurement values with a choropleth map divided into counties much of the pattern which is of interest to us would disappear.

Cesium-137, coverage of soil surface, May-October 1986 kBq/m^2

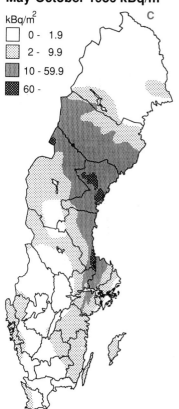

Building blocks of maps

Just as there are building blocks for charts, so too there are building blocks for maps. Some of the building blocks for maps are new, but most of them are the same as for charts.

Where the legend is put depends on the appearance of the map. For maps of Sweden it may be an advantage to place it to the left directly under the title so as to create a neat layout on the map and make it easy on the eye. To make it easier for the reader, the units of measurement should also be written in the legend.

Class boundaries, shading, patterns and boundary lines

There are no clear-cut rules for class divisions, except that it is often necessary to compromise between different aims. As with all class divisions we are aiming for classes of equal length. For statistical maps it is also desirable that there should be approximately the same number of areas in each class. Just as with tables our aim is that natural clusters of observations should not be split by a class boundary.

For certain maps the class boundaries can be predetermined by the fact that certain boundary values have legal or objective significance. For maps which show increases or decreases it is appropriate to have a "practically unchanged" class in the middle where small changes are included.

In a map which shows ratios there is a natural order between the classes. In this case it is best to use shading which goes from light to dark to visually show the order. A map which shows values on a nominal scale should instead use patterns to show that there is no ranking between the classes. In order to produce a readable map the number of classes and the accompanying patterns should be limited to at most five. With only four patterns it may already be difficult to distinguish the different areas. If you use colour you can have more classes. One colour in different shadings together with different grey shadings can be very effective (see page 85).

If a map contains lakes then it is appropriate for these to be white, in which case white should not be used as a pattern. Black is unsuitable as a pattern, partly because the areas concerned are too dominant and partly because it is difficult to obtain good print quality for large, black areas. The boundary lines between two black areas disappear too.

The appearance and number of boundary lines depends on what you want to show with the map. They should be drawn unbroken and of different thickness if the lines show boundaries on different levels.

Examples of choices in dividing classes

The maps on the right show the level of employment for the south part of Sweden in 1991. Below is the distribution of measurement values for the districts in the area.

Map D has class divisions which emphasises the extreme values for the variable.

In map E the variable is divided into three classes with an equal number of districts (11) in each class.

Map F shows the data in four classes which use the spread of the variable and a "natural" class division.

Map G has the same divisions as F but with a reversed order of shading.

We recommend the choice of map F since four classes give a better depiction than three. The class boundaries are suitable and the order of shading is suitable – higher employment is depicted with darker shading.

See chapter 3 for a full description of the building blocks of charts and the principles of, for example, writing titles

It may also be appropriate to show the scale and a North arrow if the area being shown cannot be presumed to be familiar to the reader

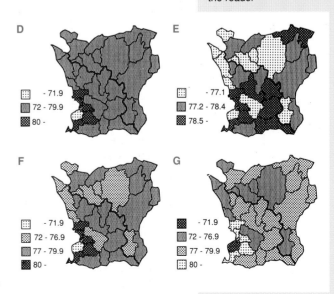

Density maps – showing numbers

Density maps are suitable for showing numbers, i.e. absolute frequencies. There are three variants of density maps: exact location density maps, object-number density maps and reference density maps. For the first two variants it is necessary to know the exact geographical location of the objects. For reference density maps it is sufficient to know how many objects exist within an area. In this chapter we will describe only two variants of *reference density maps*: density maps showing frequencies using unitary symbols and density maps showing frequencies with proportional symbols.

Uniform symbols

Map A shows the number of cars for each district in county 11. Each symbol represents 2,000 cars. The more symbols a district has, the more cars there are. However, the map says nothing about where in the district the cars are to be found. Maps with uniform symbols are easy to read.

Map B in the margin also shows the number of cars for the districts in county 11. Here the dots are randomly scattered within the districts. This gives the false impression that each dot is meant to show a number of cars in connection with the location on the map.

Proportional symbols – circles

Maps A to C shows the same set of data. A circle is drawn in the center of each district and its area is proportional to the number of cars in the district. The bigger the circle, the more cars there are in the district. However, the map says nothing about where in the district the cars are. If you think that the reader is not familiar with the different areas, you can write the names of the areas.

Map D shows the number of households in localities in county 11. The circles are placed in the centers of the respective localities.

It is important that you choose circle sizes that give some harmony between the biggest and the smallest circle. At worst you will not be able to draw this type of map if the differences are too great. Map C is good, but map D is a borderline case.

In computer programs the circles are usually drawn independently of one another. If they end up on top of one another then you must see to it that the large circles are drawn underneath. It is more difficult to read the values in maps C and D than in map A, but this type of map saves space and can tolerate a certain degree of overlapping of the circles.

Data for maps A to F

District	Cars	Lorries
E	36,058	2,892
S	12,053	1,114
V	4,387	394
K	13,891	1,247
F	7,526	641
G	3,940	215
N	21,180	2,036
T	3,964	208
O	4,704	268

In map A we have chosen to round off the number of cars to the nearest 2,000

Unsuitable map

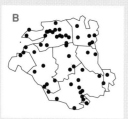

B is a reference density map but looks like an exact location density map

What distinguishes map C and map D, apart from the variable shown, is the size of the area each circle is linked to. Map C depicts districts, while map D depicts built-up areas

Cars in the districts of county 11

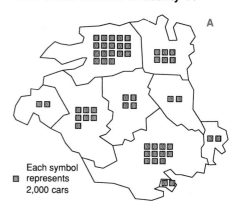

Each symbol represents 2,000 cars

Cars in the districts of county 11

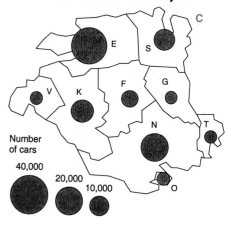

Number of cars

40,000

20,000

10,000

Households in localities of county 11

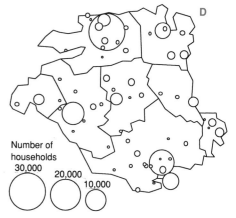

Number of households

30,000

20,000

10,000

Cartograms – a cross between charts and maps

With cartograms the map is usually regarded as a background map for the identification of the geographical area for the variables to be shown. Small charts or symbols are then drawn in the areas according to the general rules which apply for charts. It is important that the symbols are made clear and distinct to suit the printing process or copying.

Pie charts of different sizes

Map E shows how the number of cars and lorries is distributed between the districts. The circles are placed in the districts and the bigger the circle a district has, the higher the total number of vehicles.

If the circles overlap then you should put the smallest circle on top, or preferably move the circles so that they do not cover one another.

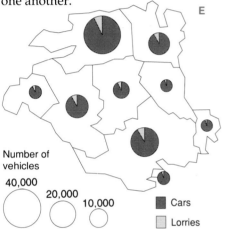

Bars of different sizes

Map F shows the number of cars for each district. The higher the bar is, the more cars there are in the district.

The bars are put in the appropriate place in the districts so that they do not overlap one another. If you cannot find enough room for the bars in the district then you can put them outside of the map and use an arrow to show where the bar belongs.

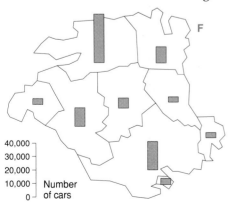

Maps with symbols

In addition to ratios and frequencies maps can also show modes. Modes can be shown either with symbols or with patterns in order to show that there is no ranking between the classes.

Map G shows the branch or branches of agricultural production which make up a higher proportion than the national average in each county.

Resist the temptation to use symbols of different size. This makes the map much more difficult to understand.

Characteristic production trends in agriculture in Sweden´s counties 1991

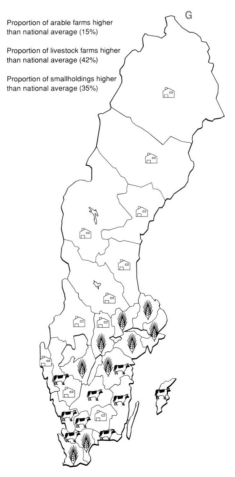

There is no generally accepted idea of what should and should not be regarded as a cartogram. Different authors have set different boundaries between statistical maps and cartograms

You may use any chart types and symbols in a cartogram. The more sophisticated a chart is, the larger the area it will require on the map

11 Some chart philosophy

The contents of charts are discussed on this and the following page, while the form of charts is discussed in the rest of the chapter. The whole book is an example of how charts are incorporated into a larger context.

How can you convey your message using charts in an effective way? To succeed you need to be aware of the functions of charts from different perspectives:
- Feeling for the chart's *contents*. You need to know what are the important results for the target group you are addressing.
- Feeling for the *shape* of the chart. You need to know how to draw different types of charts and be able to choose the appropriate type for your message and your target group. You have to be able to draw the chart so that it gives a true picture of the data, is easy for the reader to assimilate and is aesthetically pleasing.
- Feeling for the chart's *context*. Each chart will be combined with a written text or verbal presentation, tables and other charts to form a whole with a clear structure.

Up to now we have considered different types of charts and how to draw them. In this chapter we will complement this knowledge of the mechanics with some important principles which you also need to understand in order to be able to produce good charts.

Subject matter and target group

A chart should have an interesting content and be adapted to the target group. Charts are a way of presenting the results of an analysis, where the subject matter is decisive, and at the same time they are a means of statistical communication, where the target group is also decisive.

What the subject matter means for the chart

The subject matter determines how the data should be analysed and what comparisons should be made.

You want to compress data graphically to make the contents of a large table clear

You should process your data bearing in mind the subject matter: Which variables should be included? What calculations are necessary?

You want the reader to make graphical comparisons

The subject matter determines which comparisons are important. These should be clearly visible. Different parts of the subject matter may require different charts.

You want to show important patterns

Many relational and developmental patterns can only be seen in charts. You need to work a lot on your data using different analytical methods to be able to discover these patterns.

What the target group means for the chart

Your charts are only good if the target group thinks that they are interesting and can understand them.

The target group's knowledge of the subject

A target group which knows little about the subject requires charts which provide an overview, so you should draw charts using simple concepts. The readers should understand the titles and the variables.

A target group which knows a lot about the subject is prepared to study charts carefully, so your charts can contain more details and allow more comparisons.

The target group's experience of reading charts

When you are addressing readers who are not familiar with statistical charts your message must be simple and clear. Charts showing several variables at the same time, such as bar charts with stacked or grouped bars, can be difficult to understand.

If the target group is used to reading charts then it is possible to produce complicated charts with many variables. You can use advanced scales such as indices and logarithms.

An illustrative example: Is there wage discrimination against women?

In order to throw light on this subject we proceed from a set of data which refers to full-time industrial employees in 1990. Table 1 has been taken from the statistical source. First we convert into relative frequencies and compare the distributions in table 2. A clear pattern emerges: for all of the salary classes below SEK 12,000 there is a lower proportion of men than women; for all of the classes above SEK 12,000 there is a higher proportion of men. We summarise table 2 in table 3, which becomes the basis for chart A.

Table 2. Relative salary distributions by sex

Monthly salary	Men	Women	Difference
	Per cent		
-7,999	0.6	6.3	-5.6
8,000 -8,999	0.6	4.5	-3.9
• • •			
11,000-11,499	4.2	7.0	-2.8
11,500-11,999	5.3	6.5	-1.2
12,000-12,499	5.9	5.3	0.5
12,500-12,999	6.5	5.0	1.6
• • •			
25,000-	6.9	0.9	6.0
Total:	100%	100%	

Table 3. Relative salary distributions by sex

Monthly salary	Men	Women
-11,999	20.7	65.6
12,000-	79.3	34.4
Total:	100%	100%

Women at different occupational levels

Per cent

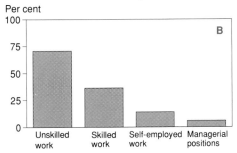

Women's salaries in relation to men's

Percentage difference

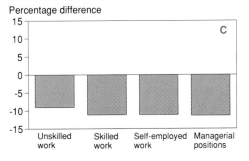

Low-paid industrial employees 1990
Earning less than 12 000 SEK per month

Per cent

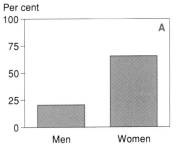

Chart A shows that women have lower salaries than men. The next step is to examine the reasons for this difference. The relationship could be indirect (arrow 1) with women having lower positions than men and therefore lower salaries, or direct (arrow 2) with women having lower salaries compared with men in the same positions.

In order to examine arrow 1 we use the information in table 4 to calculate the proportions of women at different occupational levels. The results are shown in chart B.

Relationship 2 is examined by comparing the average salaries in table 5. We use the data to calculate the percentage differences in salaries which are shown in chart C.

The subject matter can thus be expressed in three questions, each of which is answered by its own chart:
- Do women have lower salaries? (A)
- Is this because women are to be found at lower occupational levels? (B)
- Do women have lower salaries at the same occupational level? (C)

Interesting charts do not just appear on their own. You have to structure the subject matter in an interesting way, make the appropriate calculations (because data is seldom in a form suited to your situation) and choose a suitable chart type.

In this case we have chosen the simplest possible chart type and made the charts as similar as possible. By doing so we make the charts suitable for a large target group.

Table 1. Salary distribution

Monthly salary SEK	Men	Women
	number	
-7999	1070	3327
8000 -8499	947	2370
8500 -8999	1394	3602
9000 -9499	2172	4415
9500 -9999	3361	4813
10000-10499	4104	4638
10500-10999	5539	4517
11000-11499	6883	3701
11500-11999	8785	3473
12000-12499	9692	2837
12500-12999	10818	2635
13000-13499	9527	1836
13500-13999	9702	1794
14000-14499	8063	1294
14500-14999	7417	1143
15000-15499	6547	939
15500-15999	6239	881
16000-16499	5400	611
16500-16999	5295	624
17000-17499	4547	427
17500-17999	8030	731
18000-18999	6846	641
19000-19999	5557	422
20000-20999	4384	303
21000-21999	3716	219
22000-22999	3068	176
23000-23999	2736	154
24000-24999	2279	116
25000-	11484	495
Total	165602	53134

Table 4.
Employees by level and sex

Occupational level	Men	Women
Managerial position	15,380	963
Self-employed work	107,708	17,453
Skilled work	36,799	20,947
Unskilled work	5,714	13,770

After calculations:

Occupational level	Per cent women
Managerial position	5.9%
Self-employed work	14.0%
Skilled work	36.3%
Unskilled work	70.7%

Table 5. Average salaries by level and sex, SEK

Occupational level	Men	Women
Managerial position	28,630	25,376
Self-employed work	17,579	15,613
Skilled work	13,729	12,185
Unskilled work	11,664	10,597

After calculations of wage differentials:

Occupational level	Women-men
Managerial position	-11.4%
Self-employed work	-11.2%
Skilled work	-11.2%
Unskilled work	-9.1%

The wage differentials are calculated as follows:

$$100 \cdot \frac{(25,376 - 28,630)}{28,630} = -11.4$$

The design and function of charts

In this section we introduce some principles which can be used to analyze the characteristics of charts. In order to understand why a chart is good or bad you can examine the following questions, which are closely related, but which describe different aspects:

- Does the data dominate, i.e. what proportion of the chart depicts data values?
- Is the data density, i.e. the number of data values per unit of area, high or low?
- Does the chosen chart type give a clear picture of the data?
- Does the graphic picture give a true picture of the data?

In order to understand the characteristics of a chart you also need to know something about the capacity of the eye to make comparisons. We shall therefore conclude this chapter with a section on perception.

Let the data dominate

In a chart we find both building blocks which directly show data and building blocks which are there to help the reader to understand the contents of the chart. Both the building blocks for data and those for help may be suitably or not so suitably designed and are sometimes even unnecessary. Since charts are there to show data you need to be able to decide what should dominate and what is unnecessary.

Balancing the plot area with the chart area

The data set is depicted within the plot area and as the chart constructor you need to find a balance between the plot area and the chart area. Since the data should dominate, the plot area should fill as large a part of the chart area as possible, without the text outside the plot area becoming so condensed that it becomes difficult to understand.

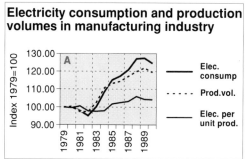

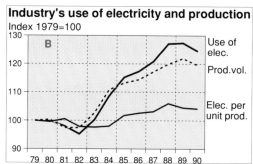

Charts A and B occupy the same chart area, about 27cm². Chart A corresponds to the basic setting in a common PC program. A revision of the chart's text makes the plot area three times bigger (the plot area in chart A is 5cm² and in B 15cm²). In chart A the text dominates, while in chart B the plot area dominates. The title text often has to be condensed, in which case the concept can be explained in detail in notes placed close to the chart.

In the plot area the data should dominate

The basic settings in the same PC program also provide grid lines and curves with symbols as in chart C. In chart B we have removed unsuitable building blocks. The number of grid lines has been reduced from 18 to eight and we have also chosen a more discreet pattern. It is unnecessary to mark the data with both lines and symbols – in chart B the interfering symbols have been removed. We have also written the legend outside the plot area.

Industry's use of electricity and production

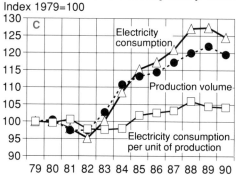

Edward Tufte: "Above all else show the data"

William Cleveland: "Make the data stand out. Avoid superfluity"

Chart area and plot area are defined on page 17

Do not accept the basic settings of your graphics program. Instead, edit the chart yourself

Certain charts in this section are unsuitable without having been given the title: **Unsuitable chart** The fact that they are unsuitable is evident from the discussion in the text

Is the printer's ink used efficiently in the plot area i.e. is it used to give a clear picture of the data? For chart B we would answer yes, because the only thing which could be removed are the grid lines. However, these discreetly drawn lines do not disturb the time series curve. Instead they contribute to the reader seeing the data clearly, such as when the changes occur and how large the gradients are. For chart C, on the other hand, the answer is no, since a lot of printer's ink is used for the grid lines, symbols and text which conceal data rather than help reveal it.

No unnecessary printer's ink for data

In the following example the same data is depicted in four charts which use the printer's ink with varying degrees of efficiency. Chart D is a common way of illustrating this type of data – a set of data with two dimensions is drawn in three. The third dimension is not used to depict any statistical characteristics. It is only there as decoration. Because it is difficult to read the height of the "pillars", this type of chart usually includes numerical values too, which makes it even more overloaded.

If the unnecessary dimension is removed about half of the printer's ink disappears and we get bar chart E. Here too it is possible to drastically reduce the amount of printer's ink by replacing the bars with lines and dots as in dot chart F. Apart from printer's ink we also save on plot area. By dispensing with the lines we ultimately get barometer chart G where the plot area contains only that which is absolutely necessary.

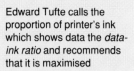

Edward Tufte calls the proportion of printer's ink which shows data the *data-ink ratio* and recommends that it is maximised

We recommend that you do not give numerical values in the plot area, but in a separate table. The only exceptions to this rule are pie charts and flow charts which do not have any scale axis

Unemployment according to age, 1991. Four alternative charts

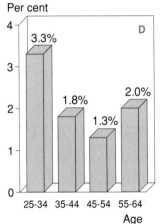

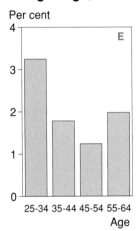

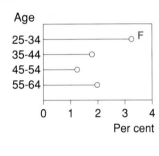

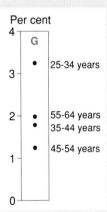

Although chart G makes the most efficient use of ink, it is not necessarily the best. Charts E and F maintain the order between the age classes so that the eye quickly grasps the relationship between age and unemployment. If your readers are unused to interpreting charts the ordinary bar chart E may be best. The extra ink in chart E is not superfluous if it increases readability and contributes to capturing the attention of the reader.

Data density

Charts E to G show four observations in two dimensions: age and unemployment. The data density of the charts can be defined as the number of values $(4 \cdot 2 = 8)$ divided by the size of the plot area. The following measurements of data density apply:

E: $8/13.4 = 0.6$ values per cm^2
F: $8/6 = 1.3$ values per cm^2
G: $8/3.3 = 2.4$ values per cm^2

Charts with a lot of detail require larger plot areas. When you wish to divide the area available in a report, you can adjust the sizes so that charts of the same type have approximately the same data density. Formerly charts were drawn by hand and they often became unnecessarily large because the hand's capacity to draw details is small compared with the eye's capacity to see them.

Data in chart A:

Low	3.3
Medium	2.9
High	1.2

Data in chart B:

Low	Men	3.3
Low	Women	3.2
Medium	Men	3.3
Medium	Women	2.4
High	Men	1.4
High	Women	1.1

Adapt the plot area to the number of values

The plot area should be adapted to the number of values. Chart A shows six values and chart B shows eighteen values. We have therefore made the plot area in chart B three times bigger so that the data density is the same size in both of the charts – 0.9 values per cm^2.

Per cent unemployed 1991

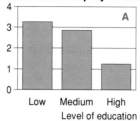

If charts A and B are to be presented together they should be of the same size and have the same scale on the y-axis so comparisons are easy.

Per cent unemployed by sex and education

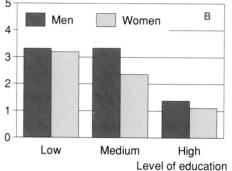

Do not make the plot area too large

You can increase the data density by increasing the number of measurement values in the chart. This can make your chart more effective and more interesting. You can also try to reduce the plot area. This may make it easier for the reader to survey the chart. Moreover, it is easier to insert a small chart into the main body of the text.

Time series charts and maps can have a very high data density. Chart C contains monthly values for eleven years, i.e. 132 two dimensional observations (time and seasonally-adjusted values). The data density is about 15 numbers per cm^2 (264/ 17.4). Chart D also contains smoothed values and the data density is about 23 numbers per cm^2. The higher data density is one reason why we recommend chart D.

A statistical map with many areas can have high data density. Map A on page 58 depicts how the proportion of those in employment varies between districts. The location of each district is depicted by two dimensions. The map contains 3 · 286 numbers and the area is about 40 cm^2. The data density is therefore 21.5 numbers per cm^2.

Order inflow, volume index, seasonally-adjusted values

Index 1980=100

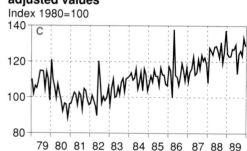

Order inflow, seasonally-adjusted and smoothed values

Index 1980=100

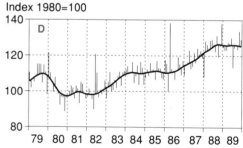

Degree of clarity

In the preceding section we discussed problems which are due to unnecessary or unsuitable graphical objects concealing data in a chart. Here we consider problems which stem from the data set and which can make charts "muddled" and difficult to interpret.

Too many bars? Choose a different chart type.

Grouped bar charts often become difficult to interpret if they have many categories in each group and if the number of groups is large. Three time series are illustrated in a muddled bar chart on page 36. This should be replaced by a line chart which gives a clear picture of the development of the time series. On pages 48 and 49 we show that dot charts can give a clear picture of relationships even if there is a large number of classes.

Jagged curves? Smooth them.

Why is chart C on the previous page so muddled? Because it contains many data values, sharp angles and lines with different gradients. All of these angles and sloping lines make the chart restless and contribute to the reader receiving an unclear picture of the development of the series. With appropriate moving averages the time series can be smoothed. In chart D there are both smoothed and seasonally-adjusted series and the two series are joined by vertical lines. Chart D is therefore a chart which simultaneously gives an overview and details – the curve gives a clear overview and the needles provide details without disturbing the picture of the whole.

Lots of curves? Draw several small charts.

A line chart becomes difficult to read if it contains many curves which often intersect one another. In chart E trying to distinguish between the series by using different patterns does not help.

By drawing a suite of charts with a small number of series in each one we can give a clear picture of the same data. All charts in such a suite should have the same size and scales on the x and y axes and contain a *common reference curve* – in this case the series for Sweden.

Charts C and D have been taken from pages 42–43

Rate of inflation in 7 countries
Per cent per year

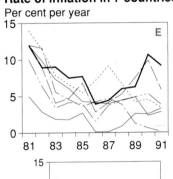

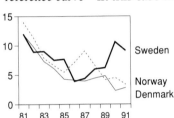

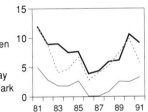

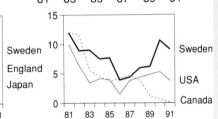

Lots of dots? Aggregate or draw several small charts.

In a scatterplot we can distinguish different groups with symbols. If these groups of points overlap as in chart F, the chart becomes unclear – it is difficult for the reader to see the relational patterns which exist.

Instead of drawing the cluttered chart F, we can either draw a suite of small charts for different parts of the data (same scales, common reference curve), or we can draw a chart for aggregated data which depicts the relationship between x and y averages. The advantage with a suite of small charts is that we show both an overview and details – no information is lost.

Food as a proportion af consumption

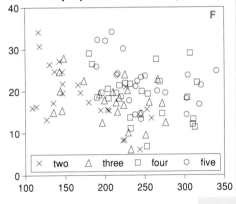

This example is discussed on page 47

Several small charts

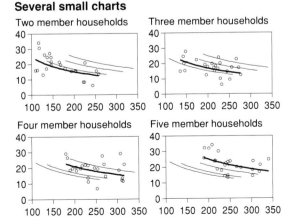

Aggregated data

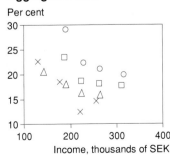

Income, thousands of SEK

The disadvantage with showing only aggregated data is that information about variation is lost.

SEK = Swedish crowns

True comparisons

A chart gives a graphical picture of data. This picture should both attract the attention of the reader and convey the correct statistical information. Often efforts to "popularise" information result in the chart being exaggerated and misleading. In this section we discuss some chart types which can give a false picture of data.

Pictograms

Charts similar to chart C are a common alternative to bar charts. The height ratio in chart C is 1:2.2 (157:350), but the chart gives a misleading comparison because the reader perceives the area of the cars and the area ratio is about 1:5. Chart B is a better alternative which both attracts attention and gives a true picture of the data because the columns of cars are of equal width. Therefore chart B gives the same picture of the data as chart A. In pictograms the numerical values should be given.

Car sales plummet!

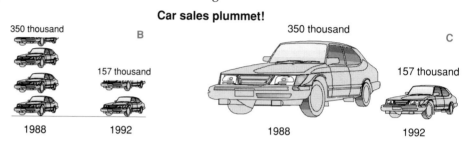

Bars and the zero on the y-axis

With line charts and dot charts the zero does not need to be included on the y-axis. This is discussed on pages 39 and 46. But in bar charts where the bars dominate so strongly, it is best to include the zero on the y-axis. In chart E the reader could be given the impression that the men's hourly wages are twice as high as women's.

Charts with a false third dimension

With PC technology it is now easy to make three dimensional charts and so they have become a common sight. These charts often give misleading comparisons. When data only has two dimensions we speak of false three-dimensionality.

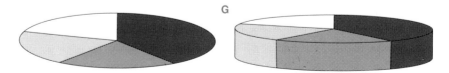

In chart F both the angles and the sector areas give correct pictures of the data. The charts in G are two variations of three dimensional pie charts. The charts in G do not give true comparisons, since the sectors which in fact are of the same size have different angles and different areas.

Tufte calls the ratio 5/2.2 the chart's lie factor. You show 5 when you mean 2.2

Chart C actually shows volymes and the volume ratio is 1:11

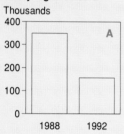

Newly registred cars

Thousands

SNI 31: Food industry
SNI 342: Printing industry
SNI 35: Chemicals industry
SNI 38: Engineering industry

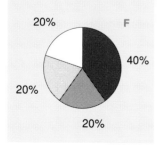

Three dimensional charts have become popular for two reasons. The charts are perceived as attractive, we get depth in the picture and our actual vision is of course three dimensional. Furthermore these charts symbolise the triumph of data technology – what is almost impossible to draw by hand is now easy to produce.

Apart from three dimensional pie charts, bar charts with a false third dimension are also common. Chart H is relatively harmless since the four pillars give a true picture of the data even if the third dimension does obstruct the reading of the heights of the pillars.

We get chart J by rotating chart H. In this way we produce perspective effects which can be confusing. The readers unconsciously expect that pillars of equal size will look smaller the further away they are. Notice that PC programs do not draw correct perspective – pillars of equal size are drawn the same size wherever they are.

Charts with a genuine third dimension

In earlier chapters, when a set of data consists of three variables we have drawn two dimensional charts. Two variables have been depicted by the x and y axes and the third with area patterns, line patterns or symbols. Chart K illustrates the following relationships:

The two explanatory variables are depicted by the axes on the base plane of the chart and the dependent variable is depicted by the vertical axis. Therefore all three dimensions in the chart are meaningful.

Chart K has two disadvantages: It is very difficult to read off the heights of the pillars which are far from the "walls" and the data needs to be structured so that the pillars do not obscure one another.

Likewise in chart L the set of data is three dimensional: time, type of oil and quantity delivered. Despite the fact that the three dimensions are real, the chart is quite useless for comparing the levels of the time series. Which of the three series had the highest value in 1984?

Unemployment according to age 1991

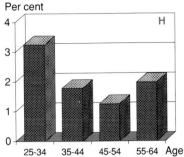

Unemployment according to age 1991

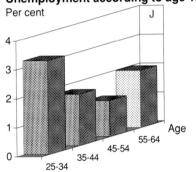

Unemployment according to age and education 1991

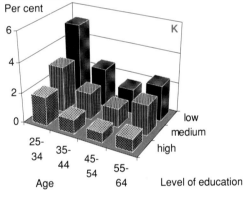

Deliveries of oil to end users

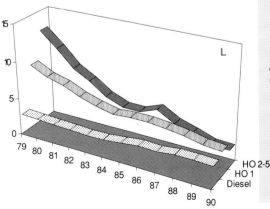

This page shows the difficulties of producing a neat page layout with three dimensional charts. Charts J to L with slanting axes are difficult to combine with right-angled blocks of text

Diagram L is a three dimensional variation of the line chart on page 36

About perception

The eye's capacity to make comparisons

What sorts of comparisons can the eye make with good accuracy and what comparisons are difficult? William Cleveland has made a compilation of experiments where the subjects had to make different graphical comparisons. Here we describe the most important results, beginning with the cases which are easiest for the eye.

Positions on the same scale

When the eye compares the lines in chart A the position of the end points on the x-axis is estimated. The level of the curve in chart B is interpreted by the eye comparing the positions of the four values on the y-axis. Grid lines (vertical in chart A, horizontal in chart B) would make it easier for the eye.

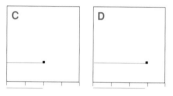

Position on several identical scales

When the eye compares the lines in charts C and D the positions of the end points on the two x-axes with identical scales is estimated. Vertical grid lines would make it easier to make judgements.

Lengths

When the eye compares the bars in chart E the lengths are compared by estimating the positions of the four end points on the x-axes simultaneously. Vertical grid lines would make it easier for the eye.

Gradients and angles

In order to estimate how the gradient of the curve in chart F changes, the eye compares the three angles marked. Horizontal grid lines make this easier. It is more difficult to judge the angles in chart G, since the sides of the angles have different orientations.

Areas

Estimating the size of areas is difficult. Which of the areas in histogram H is the largest? How much larger is it?

The correct answer is that area 2 is about 40% bigger than area 3. The sector areas in chart G are also difficult to estimate.

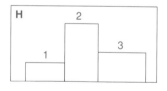

Volumes, shading and colour intensity

The eye can rank volumes (if they have the same shape) and areas with different degrees of shading or intensity of colour, but making quantitative estimates of volumes or shading is almost impossible. Volume 1 is about 6.5 times bigger than volume 2. Areas 1, 2 and 3 have 20%, 30% and 60% shading.

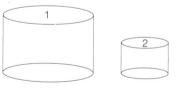

Easiest

Easier

Easy

How easy is it for the eye to make comparisons?

Difficult

More difficult

Most difficult

Make it easier for the reader

Often you can choose between different types of chart for the same set of data. In this case you should choose the chart type that makes it easy for the reader to make the most important comparisons.

In pie chart J, the reader must to be able to judge areas and angles. If instead you illustrated the same data using stacked bars as in chart K, you would only require the reader to compare lengths and positions on the scales.

If you want to show how the unemployment varies between districts, you can draw this in different ways. Chart L requires the most of the reader, i.e. that the eye can distinguish between shadings. Chart M requires the eye to compare lengths (heights of bars), and chart N makes it even easier because the frames act as identical scales.

Optical illusions

There are many optical illusions which can deceive the eye. Histogram H on the previous page is an example of an optical illusion which reduces the height of bar 2 because it is divided in the middle by a horizontal line. Chart O illustrates the illusion of the dividing line.

The eye sees at right angles

The differences between curves are almost impossible to interpret. This is a problem with both area charts and line charts. The intended comparison is in the y-direction, i.e. vertical, but the eye takes in the shortest distance irrespective of the angle. When the curves slope strongly upwards or downwards the eye is completely fooled.

Area chart P depicts a total which consists of two parts. Part 2 is the difference between the upper curve and the curve for part 1. The eye does not see that part 2 is constant between years 70 to 77 and that afterwards the series follow each other.

In line chart Q it is clearly seen that part 2 is constant between 70 and 77, but it is difficult to see that the difference between the parts is constant afterwards. The eye believes that the curves are closer together between 79 and 80, and further apart between 81 and 85. If you want to show the difference between the curves you have to show this in a separate chart like R.

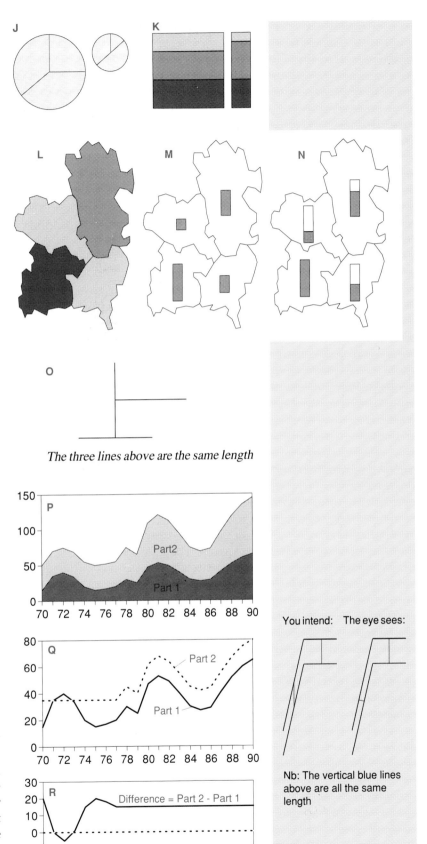

The three lines above are the same length

You intend: The eye sees:

Nb: The vertical blue lines above are all the same length

12 Charts and layout

What is a layout? It is a graphical composition. Layout is the design and relative positioning of the main body of text, titles, charts, maps, tables and pictures.

Try to make your layout as clean and simple as possible. As we have already pointed out it is the context of the text and pictures which should attract attention and interest. At the same time it is important to create a structure which gives a natural relationship between the pictures, text and tables. The reader should not have to search for commentaries or illustrations. A good layout should catch the reader's eye, encourage him or her to read and help the reader to find the way through the pages.

Charts may be used for different purposes, such as conveying main findings in a more pictorial way or attracting attention. The layout is influenced by these aims. In sections with a lot of tables and text, for example, you may be justified in using charts to make the section less heavy.

The overall structure

Right from the start you need to be clear about whether the reader will be seeing one page at a time, as in a duplicated report, or a complete double-spread of pages, as in a book. Both the problems and possibilities are considerably greater in the latter case, which is why we will concentrate on it.

Charts and text

Your task is to create a natural reading order for the reader. A useful rule is that introductory paragraphs should be used to present the subject matter and the concepts which form the background to the charts. Then you should put the chart itself and follow it with any comments about the chart. Vary the space between the text and charts so that it is immediately clear which text belongs with which chart. Avoid putting charts and their accompanying text on different pages or spreads of pages.

Tables

After the chart has attracted attention and given an overview you can continue with a table which provides more detailed information and the possibility of continued analysis and detailed commentary. Tables which are not discussed in the text, but on which the chart is based, may be placed in an appendix. This also applies to very large tables which would otherwise split up the main body of text and create unwanted interruptions.

Make a synopsis

Working with a synopsis means that you make a clear summary or overview of the *whole* of the contents in outline form. Feel free to sketch the different pages by hand first. This saves you a lot of time because you soon see if the layout is suitable. Working with a synopsis gives you better control over the layout. You can easily try out different solutions, for example choosing between one or two columns (see next page) in order to make the whole thing interesting and clear. Finally, the use of a synopsis facilitates the arrangement of a book or report, by obliging you to decide how many pages a section will contain early on.

Another reason to use a synopsis is to create better discipline in your handling of the text. You simply get no space to be longwinded

Working on double-spreads

Work with pages spread by spread. You always see a double-spread of pages in a book and so you have to work with a total layout for the whole of the double-spread. By working with spreads it is easier to achieve a good variation and balance on the pages.

An appropriate mix of charts, text, tables and empty space creates a harmonious and interesting double-spread. Think about making things easier for the reader.

Examples of roman fonts:
Times,
Palatino,
New Century Schoolbook

Examples of grotesque fonts:
Helvetica,
Futura,
Avant Garde

Roman fonts (serif)

Roman fonts use heels, strokes and swellings to give a coherent and legible word formation. They are legible and well suited to continuous text and appear agreeable and good for long texts.

Grotesque fonts (sanserif)

Grotesque (upright) fonts appear cold, rational, pushy and impersonal. They are very suitable for titles, and text accompanying pictures, diagrams and tables. It also works in short texts.

The tools of layouts

Fonts

The choice of font and size is just as important as the choice of page format and the number of columns. Two different fonts are usually enough: a roman font for the main body of text and a grotesque font for the titles. For the main body of text it is best to use a roman font such as Times or Palatino. It makes the continuous text readable and takes up less space than a grotesque font.

The titles in the text should be in a different font, possibly a grotesque font such as Helvetica or Universe, and in bold type. Do not use too many levels of titles – try to manage with three, or at most four. For the lowest level it is usually best to choose the same font as the main body of text, but in bold, as we have done on this page.

Titles of chart and tables should also be in a grotesque font with bold type and placed above the chart or table. Text in charts and tables should also be in a grotesque font. This means that you can quickly distinguish between diagram text and the continuous text.

One or several columns

If the publication is in a large format it is a good idea to divide the text area into two or more columns. With several columns it is easier to vary text and pictures and to achieve a flexible layout. Diagrams can cover two columns if they are complex or contain a lot of data. If possible, they should then be put at the bottom of the page in order to create a good balance. It is good to make simple diagrams small so that they fit into the column.

A constant single column is only used in small, simple publications, when the main body of text is short or where there are lots of large tables.

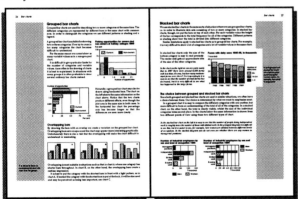
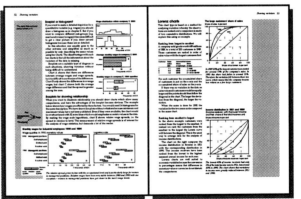

How we made this book

We have chosen to use Palatino for the main body of text. We have then alternated quite freely between one and two columns. We have chosen Avant Garde for the titles, while for the text in the margins, tables and diagrams we have used Helvetica. Straight commentaries on diagrams have been written in italics so that they stand out from the normal continuous text. In certain cases the diagram has been allowed to extend out into the margin, which is otherwise used for comments, explanations and examples. As you will no doubt have noticed, we have chosen to work with very "active" margins.

13 Charts in practice

In this chapter we shall apply the principles given in the earlier parts of the book to some realistic examples. For each example we begin by presenting the data and a chart which will be the starting point of our discussion

Patterns and order in bar charts

A bar chart may be drawn in several ways for the same sets of data. Even if this chart type may be considered simple, you as the chart designer have to be aware of the fact that there are various alternatives and make a conscious choice.

The following data set refers to average monthly wages for industrial employees. We want to depict wages for men and women at occupational levels.

Average monthly wages for full-time industrial employees 1990, SEK

Occupational level:	Men	Women
Managerial position	28,630	25,376
Self-employed work	17,579	15,613
Skilled work	13,729	12,185
Unskilled work	11,664	10,597

Source: Wages and employment in the private sector 1990, Statistics Sweden

Chart A shows our first suggestion, a chart with grouped bars. Other alternatives for this data are horizontal bars, dot charts and barometer charts. Here we have chosen the simplest chart type which is suitable for giving a clear depiction with an easily read chart.

Charts A to C show that black and white create very strong contrasts and that black totally dominates. Chart B is dominated by the rising pattern and chart C by the falling pattern. In chart D we have made a more suitable choice of area patterns – as soft a contrast as possible, but still making a clear distinction between the categories.

The charts E and F show the effect of varying the order in which the categories are placed. The data contains two patterns: the wage differences between the sexes and between the occupational levels. These two patterns coincide in chart E, but can be distinguished in chart F. Chart G is therefore our final recommendation.

Charts B to D depict fictional data

Wages for industrial employees 1990

Thousands of SEK per month

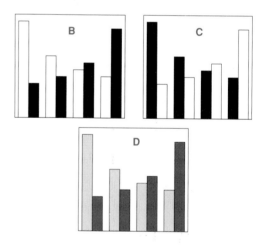

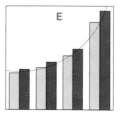

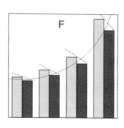

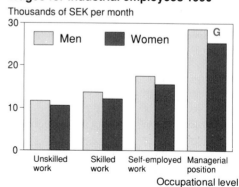

Wages for industrial employees 1990

Thousands of SEK per month

Bar chart with extra dimension

In chapter 1 we showed a chart which depicted boys' and girls' choices of subject in upper secondary school. Here we discuss that chart and some alternatives.

Intake into three-year courses at upper secondary school, autumn 1991

	No. entrants	Prop. of women,%
Theoretical subjects:		
Humanities	4,104	85
Social sciences	12,048	71
Science	8,520	52
Economics	12,431	56
Engineering	10,930	19
Vocational subjects		
Construction	465	1
Electrical engineering	815	4
Vehicles and transport	1,126	4
Business studies	1,200	62
Industrial engineering	1,752	3
Agriculture	1,289	50
Nursing (care)	3,427	87
Catering	513	58
Nursing	772	92

Source: Upper Secondary School 1991/92. Statistical report U 50 SM 9201, Statistics Sweden

Chart H is powerful, but does not entirely follow the rules shown earlier. In the first place the y-axis is a qualitative variable, school subject. According to chapter 4 this should be indicated by spaces between the bars. In the second place, half of the chart is unnecessary since the proportion of boys and the proportion of girls describes the same phenomenon. We should therefore be able to halve the plot area (i.e. double the data density) by removing half the chart.

A strict application of earlier chapters would give chart J or K or a dot chart similar to K. The advice given earlier are basic rules, and although they give good charts, it may sometimes be appropriate to deviate from them. The slanting stack of bars in chart H gives a suggestive picture of imbalance which fits well with the subject matter (sex differences in choice of subject) and grabs the reader's attention.

But an important dimension is lacking in charts H to K – all the school subjects have the same bar width irrespective of the number of pupils! Because the subjects vary greatly in size this may give an inaccurate picture of the imbalance between the sexes. Therefore in chart L we have drawn the bars so that the width is proportional to the number of pupils. It is now evident that social sciences and engineering are major sources of imbalance.

Intake by sex into three-year courses at upper secondary school 1991

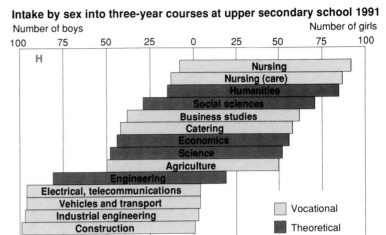

Intake into three-year courses at upper secondary school 1991

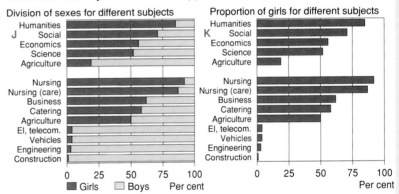

Intake by sex into three-year courses at upper secondary school 1991

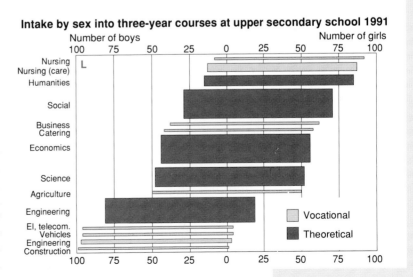

Life expectancy and bar length – the zero on the y-axis

We have taken chart A from an article in a newspaper concerning the EC's population profile. What we mainly wish to discuss are two characteristics of Dagens Nyheter's chart: the names of the countries are written on a slant and the y-axis begins at 65 years rather than at zero. Both of these characteristics conflict with the principles we discussed in chapters 4 and 11 and we shall examine how this affects the charts readability.

Another objection to the chart is that information about Sweden is not given. Comparisons with Swedish conditions are doubtless of interest to the readership of Dagens Nyheter and we have therefore added information about Sweden to the data.

Life expectancy at birth 1985-1988

Country	Men	Women
Belgium	70.0	76.8
Denmark	71.8	77.6
West Germany	71.8	78.4
Greece	72.6	77.6
Spain	73.1	79.6
France	71.8	80.0
Ireland	71.0	76.7
Italy	72.6	79.1
Luxembourg	70.6	77.9
Netherlands	72.2	78.9
Portugal	70.6	77.7
U.K.	71.7	77.5
Sweden (85-89)	74.2	80.1

Source: A Social Portrait of Europa, Eurostat 1991, and Statistics Sweden

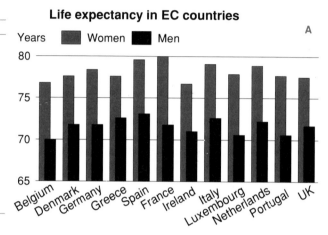

Life expectancy in EC countries

Source: Dagens Nyheter, April 30, 1993

In order to avoid writing country names on a slant, we draw a chart with horizontal bars. We also let the bars start at zero. In addition, in chart B we have made the foremost of the two overlapping bars for each country a light shade and the bar behind a lighter shade as recommended on page 26. Overlapping bars are quite suitable in this case because women have a higher life expectancy in all the countries being compared. We have placed the reference example, Sweden, at the top.

In order to ease comparisons, the countries should be ranked in the way discussed on page 25. In chart C we have ranked them according to women's life expectancies and in chart D according to men's. Of these two we would choose D because it is more difficult to compare the short bars in C than the long bars in D.

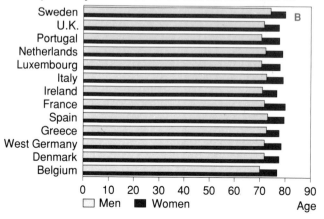

Life expectancy in EC countries and Sweden 1985-88

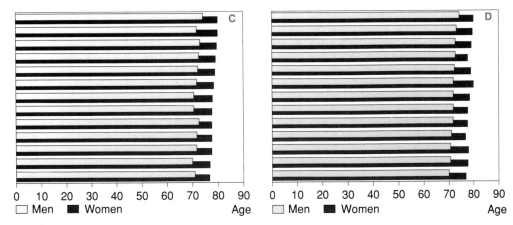

Charts B to D show that it is difficult to discern important differences. The scale on the age axis is far too cramped when the zero point is included. Only 1/9th of the plot area is used to show variations in the data because all of the variations are in the interval between 70 and 80 years. Therefore, in chart E we let the bars begin at 65 years as in the newspaper's chart. Admittedly chart E contains a graphical distortion, because the women's bars are twice as long as the men's, but because newspaper readers should know that women do not have twice the life expectancy, chart E is hardly misleading. The distortion is less problematic in dot chart F where the bars are replaced with lines.

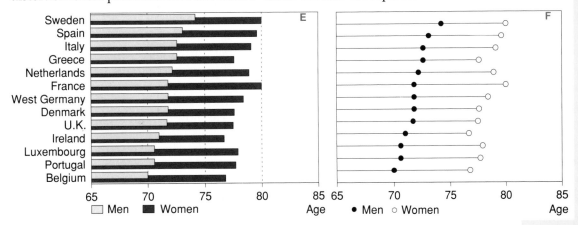

William Cleveland has a solution to this problem – when the zero is not included on the axis, the lines in diagram F are replaced with broken grid lines. The best diagram for an audience used to reading diagrams is G, but E is the most suitable for use in newspapers.

Life expectancy in EC countries and Sweden 1985-88

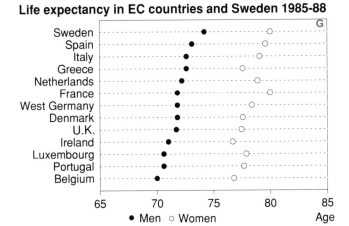

Time series on different levels – GDP in three countries

In this example we discuss the problems which arise when series on quite different levels are to be compared. In chapter 6 various alternative charts are described, such as original scale charts, index charts and semilogarithmic charts. Here we choose to test these different types of chart for a comparison of the GDPs of Sweden, Turkey and the U.S.A.

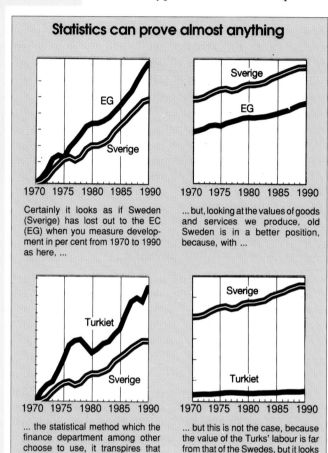

Statistics can prove almost anything

1970 1975 1980 1985 1990

Certainly it looks as if Sweden (Sverige) has lost out to the EC (EG) when you measure development in per cent from 1970 to 1990 as here, ...

1970 1975 1980 1985 1990

... but, looking at the values of goods and services we produce, old Sweden is in a better position, because, with ...

1970 1975 1980 1985 1990

... the statistical method which the finance department among other choose to use, it transpires that Sweden lagged behind Turkiet, ...

1970 1975 1980 1985 1990

... but this is not the case, because the value of the Turks' labour is far from that of the Swedes, but it looks impressive in the statistics.

Source: Dagens Nyheter February 19, 1993

Index – simple to do, hard to understand

We have taken the charts on the left from a daily newspaper. The charts lack scales, but charts 1 and 3 are index charts which show *relative* changes from 1970 on. Charts 2 and 4 are on ordinary scales, in this case US dollars per head of population, and show levels and *absolute* changes.

Unfortunately it is common for indexes to also be interpreted as if they show differences in levels and then the concept of statistics is put to shame. It is not true that statistics can prove almost anything. Rather, it is the way statistics is used and presented which can be misleading. An audience not used to statistics presented with a chart about a difficult subject will misunderstand the message unless the difficulties are explained.

How do we make true comparisons in this case with economic data from three countries of different size and at different levels of development? The important thing here is to show development in fixed prices.

Columns 1 to 3 in the table below include GDPs in national currencies, i.e. dollars, kronor and lira. To be able to draw the series in the same chart we must calculate in the same units. The simplest method is to convert to an index – the advantage being that we then avoid difficult currency comparisons.

Gross domestic product in 1985 prices. U.S.A., Sweden and Turkey, 1970-1990

	GDP, billions in national currency			GDP, billions of dollars		Populations, millions			GDP in dollars per head of population		
	U.S.A.	Sweden	Turkey	Sweden	Turkey	U.S.A.	Sweden	Turkey	U.S.A.	Sweden	Turkey
	(1)	(2)	(3)	(4)	(5)	(6)	(7)	(8)	(9)	(10)	(11)
1970	2714.4	651.73	13260	81.773	57.154	205.1	8.043	35.61	13.238	10.167	1.6052
1971	2791.8	657.89	14461	82.545	62.334	207.7	8.098	36.55	13.444	10.193	1.7052
1972	2934.4	672.95	15412	84.435	66.430	209.9	8.122	37.50	13.980	10.396	1.7714
1973	3086.6	699.65	16093	87.785	69.364	211.9	8.137	38.45	14.566	10.788	1.8040
1974	3069.4	722.02	17463	90.593	75.271	213.9	8.160	39.40	14.353	11.102	1.9105
1975	3043.5	740.46	19017	92.905	81.968	216.0	8.192	40.35	14.092	11.341	2.0315
1976	3193.8	748.29	20669	93.888	89.091	218.0	8.222	40.93	14.648	11.419	2.1769
1977	3336.4	736.35	21564	92.390	92.950	220.2	8.251	41.84	15.149	11.197	2.2218
1978	3490.0	749.24	22178	94.008	95.595	222.6	8.275	42.77	15.679	11.360	2.2349
1979	3579.2	778.01	21987	97.618	94.770	225.1	8.294	43.74	15.904	11.770	2.1666
1980	3563.8	791.00	21819	99.247	94.047	227.8	8.311	44.74	15.647	11.942	2.1022
1981	3632.9	791.26	22769	99.280	98.144	230.1	8.324	45.86	15.786	11.927	2.1399
1982	3551.8	800.05	23906	100.382	103.044	232.5	8.327	47.02	15.275	12.055	2.1915
1983	3675.0	814.54	24791	102.201	106.859	234.8	8.329	48.21	15.652	12.271	2.2168
1984	3900.7	846.98	26214	106.271	112.991	237.0	8.337	49.42	16.458	12.747	2.2863
1985	4016.6	865.79	27552	108.631	118.758	239.3	8.350	50.66	16.786	13.010	2.3440
1986	4119.6	884.99	29842	111.041	128.630	241.6	8.370	51.63	17.050	13.267	2.4914
1987	4243.3	910.20	32049	114.204	138.144	243.9	8.398	52.75	17.395	13.599	2.6190
1988	4410.6	930.83	33298	116.792	143.528	246.3	8.436	53.97	17.907	13.844	2.6594
1989	4525.3	952.88	33681	119.558	145.176	248.8	8.493	55.26	18.190	14.077	2.6274
1990	4555.0	959.95	36786	120.446	158.559	249.9	8.559	56.47	18.225	14.072	2.8077

Nb: Colums 4, 5, 10 and 11 have been converted using 1985 purchasing power parities   *Source:* National Accounts, volume 1. OECD 1993

Chart A below shows columns 1 to 3 converted to indices. Chart A is technically quite correct and is based on a reliable source, but is the chart correct in terms of contents? Columns 6 to 8 in the table show that the countries have very different population growths. If the 1990 and 1970 populations are compared, the U.S.A. grew by 22%, Sweden by 6% and Turkey by 59%. Since the growth in population is significant in comparing the countries' economic development, chart A is unsuitable. We have therefore based chart B on the index of GDP *per head of population* for the three countries. For the U.S.A., for example, the index consists of the quotient between columns 1 and 6.

GDP in fixed prices for three countries 1970-90

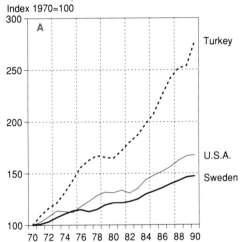

The reader is struck by Turkey's strong growth (177%) and that Sweden is "worst". The fact that the differences are due to different population growth is not shown.

GDP per head in fixed prices 1970-90

Turkey's growth is not quite as strong (75%). The difference in growth between the U.S.A. and Sweden has disappeared.

In all the charts in this example we have used the same line pattern for each country. Remember to choose unitary patterns when you draw several charts for the same set of data in a report

Choice of base year in index charts

When as the chart constructor you choose a base year, you often determine which series shall lie at the top. Thus you can influence your readers to draw incorrect conclusions through level differences in the chart which are quite irrelevant in terms of content.

Charts B, C and D show the base year's significance for the impression the index chart gives. 1975 was a recessionary year in the U.S.A., but a boom year in Sweden, while the conditions in 1978 were reversed. Since growth from a low point in the economic cycle is stronger than from a high point, the U.S.A. series is highest in chart C, but lowest in chart D. Charts B, C and D are all technically correct, but C and D are less suitable in content terms since in each case the index comparisons are made with an extreme year as the base year.

GDP per head in fixed prices 1975-90

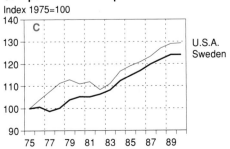

GDP per head in fixed prices 1978-90

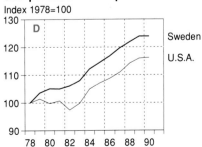

Charts for showing differences in levels

In charts A to D on the previous page we converted to indices in order to compare percentage growth in the countries' GNPs. If we want to make comparisons of levels and to compare absolute growth, we must convert to the same currency. Since currencies can be overvalued or undervalued, the conversions and hence the chart can be misleading.

GDP, 1985 prices and parities

Thousands of dollars per head

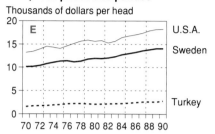

GDP, 1985 prices and parities

Thousands of dollars per head
Logarithmic scale

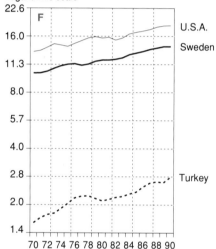

Which year's exchange rate (and prices) should be used for conversion? In certain years the Swedish krona (SEK) may be valued incorrectly in relation to the dollar. The differences in levels in the chart would then be wrong. We avoid this by using purchasing power parities. The exchange rate was 8.60 SEK/$ in 1985, while the purchasing power parity was 7.97 SEK/$. The krona was therefore undervalued in that year.

Chart E is not very good because the y-scale is so cramped that the reader only gets a rough picture of the changes in the series. This can be taken care of in various ways:
- With a semilogarithmic chart (chart F).
- By enlarging the y-scale (chart G).
- By creating several charts with different y-scales. Chart H is such a chart, where the scale is suited to Turkey's GDP.

An advantage with the semilogarithmic chart F is that we can show that the U.S.A. and Sweden have the same percentage growth and that Turkey has stronger growth without having to choose a base year. In addition the reader gets information about differences in levels. The disadvantages are that there can be no zero on the y-axis and that it is difficult to read off values between the given scale values. We prefer to describe this data with a suite of three charts:
- Chart G shows differences in levels and gives a good depiction of absolute changes for the U.S.A. and Sweden.
- Correspondingly, chart H gives a good depiction for Turkey.
- Index chart J gives a good picture of the relative changes.

GDP, 1985 prices and parities

Thousands of dollars per head

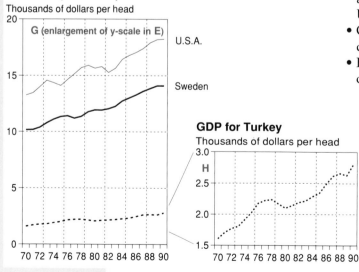

GDP for Turkey

Thousands of dollars per head

GDP per head in fixed prices 1979-90

Index 1970=100

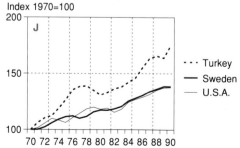

Periodical data - a chart paradox

According to chapter 6 periodical data should be illustrated by charts with seasonally-adjusted and smoothed series. Both to draw and interpret charts with periodical data therefore requires some knowledge about time series analysis. This section discusses some commonly occurring chart types, where adjusting and smoothing are not used.

Periodical data is dealt with on pages 42 – 43

Order inflow for Company Ltd 1992-93

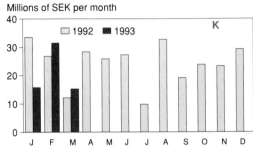

This is a common way to depict monthly data. When did something important occur? The values for 1992 do not give any information about the trend. The three comparisons for 92/93 give no clear information about increases or reductions.

Employees

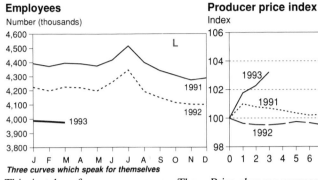

Three curves which speak for themselves

This is taken from a newspaper. The commentary is dubious. A reduction has taken place, but when? The reader is given the impression that the reduction took place at the end of each year.

Producer price index

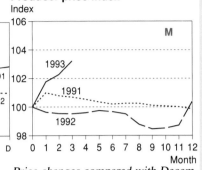

Price changes compared with December values of previous years are shown. We see what has happened and when. The effect of the SEK's declining exchange rate from Nov. 92 is obvious.

Roughly the same principle for showing periodical data has been used in the three charts above. Why does the principle work for chart M, but not for charts K and L? The answer is that charts K and L, unlike M, contain series with strong seasonal variations which make comparisons within years difficult to interpret.

Charts K and L are designed so that the reader should make comparisons in the *vertical direction*, i.e. compare one month with the corresponding month of previous years. These types of chart can only contain data for a few years and therefore do not give a picture of the trends and economic patterns of the time series. Charts of type M are suitable for depicting the short-term development of *price indices*. The reader can make comparisons within a year and also compare the rate of change between different years, because the series have been divided up into three sub-series.

The time axis in charts K to M have a scale of around 2 months/centimetre. The effect of this is that increases and reductions are almost completely hidden. Charts N to Q have around 20 months/centimetre so gradients are shown ten times more clearly. These charts depict longer time periods and give information *both* about the whole and details.

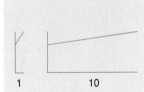

The effect of varying the length of the time axis is discussed on page 38

New orders, adjusted and smoothed

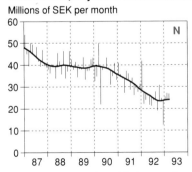

Unlike chart K, chart N gives a picture of the series' trend - the fall from 1990 to 92 has stopped. In addition N shows details of individual months, e.g. the low value for December 1992.

Employment according to LFS

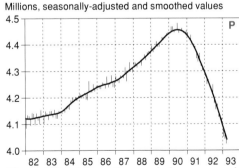

Through the longer time period we can make comparisons with the 1982 recession. The turning point in 1990 is clearly seen. The sharp decline is still continuing in March 1993.

Producer price index

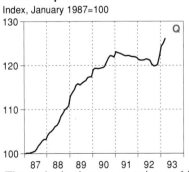

The reader is given an overview and it is possible to distinguish periods with different rates of change. The rise from November 1992 is clearly seen.

Situations with many comparisons

In this example we discuss a case with many variables and a large number of classes for each variable, which means that there are many comparisons. One chart on its own would be muddled, so it is appropriate to draw a system with several small charts. The data in this example consists of information for four years about the number of people employed divided by county and industry sector. The table below shows only part of the data.

Employed persons in different industry sectors and counties

County:	1960	1970	1980	1990	Percentage change			Difference against Sweden		
					60-70	70-80	80-90	60-70	70-80	80-90
County 1										
ISIC 1-5	57486	42895	44568	40574	-25.38	3.90	-8.96	-14.92	12.04	-2.88
ISIC 6-8	22403	22888	30735	35454	2.16	34.28	15.35	-15.05	10.75	-2.97
ISIC 9	21260	27751	45611	58762	30.53	64.36	28.83	-5.84	3.60	2.18
...										
County 24										
ISIC 1-5	35813	35816	32863	29431	0.01	-8.24	-10.44	10.47	-0.10	-4.36
ISIC 6-8	11822	11971	13848	16048	1.26	15.68	15.88	-15.95	-7.86	-2.44
ISIC 9	11936	15373	24014	30020	28.80	56.21	25.01	-7.57	-4.55	-1.64
SWEDEN										
ISIC 1-5	1843262	1650443	1516031	1423866	-10.46	-8.14	-6.08			
ISIC 6-8	772922	905965	1119204	1324275	17.21	23.54	18.32			
ISIC 9	627900	856260	1376500	1743352	36.37	60.76	26.65			

We have grouped the different economic sectors into three classes (Manufacturing etc., Private services and Public services) in order not to overburden the chart. This kind of grouping is made bearing in mind what it is interesting to show. The tables in the source contain only the number of people in employment, and to make the chart successful we must find interesting ratios. We have chosen to calculate percentage changes over a ten-year period and to compare each county's change with that of Sweden.

Our first alternative is a system of three small charts with one chart for each ten-year period. When we drew the three charts below we were confronted with two practical problems. In cases where the points ended up on top of each other we have changed the figures slightly so that all the points are visible. In addition, rather than adapt the scales to the extreme value 41, we have chosen to give this value in a footnote below the chart.

Changes in employment in different industry sectors and counties compared with totals of Sweden

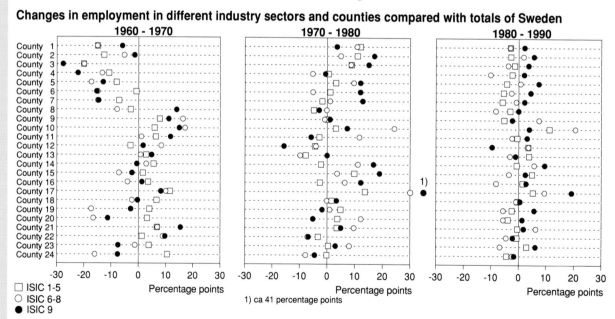

1) ca 41 percentage points

As well as this chart is difficult to read, it is impossible to apprehend the geographical variations. In order to depict these we choose to create a system with nine choropleth maps. In the chart on the next page we also use colour to give a detailed depiction of the changes. The grey colour scale depicts negative values and the blue scale depicts positive values. White is a natural area pattern for small changes.

Sidebar:

System of small charts:

Variable no. 1

Variable no. 2

Each sub-chart contains further variables which define the x and y axes. In addition, you can introduce even more variables with symbols or patterns

ISIC 1-5: Manufacturing etc
ISIC 6-8: Private services
ISIC 9: Public services etc

Changes in employment in different industry sectors and counties compared with totals of Sweden for 1960, 1970, 1980 and 1990

Difference between the county's change and the change for the totals of Sweden, percentage points

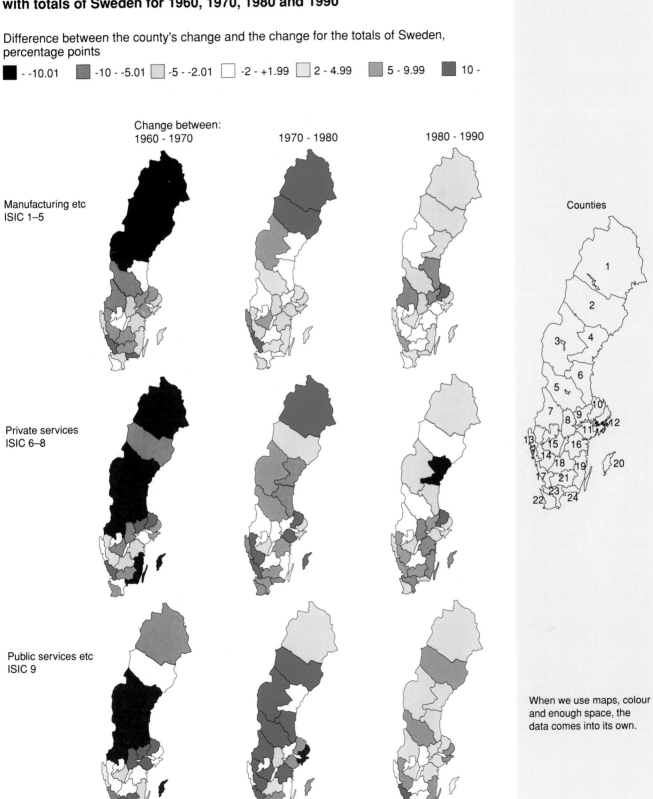

When we use maps, colour and enough space, the data comes into its own.

Pictograms and three dimensional charts are also discussed on pages 70–71

The charts on this page have been taken from Tufte's book *The Visual Display of Quantitative Information*, where they are examples of unsuitable charts

News graphics – how amusing can you be?

Just boring bars – can't we do something more sensational? This is a very common attitude towards drawing charts. Three dimensional charts and pictograms are usually the result of such an attitude.

When is it appropriate to draw charts to attract attention? The target group and the context will decide this. If, for example, you want to report on sales development to the management of a company, the fact that the target group is both interested in and familiar with the subject is important. The contents of the statistical message should then dominate the presentation, while graphical "tricks" would interfere and should be avoided. If instead you want to present statistics in a daily newspaper, you have to attract attention in order to get readers. In this situation attractive graphics which quickly introduce the reader to the subject are appropriate.

Good news graphics?

Charts A to C are three examples of news graphics. This is good if:
- The picture attracts attention.
- The picture has a clear association with the subject.
- The picture gives a good depiction of the statistical data.

Chart A: Empty barrels?

Both the symbolism and representation of data is poor. Barrels are associated with *volumes*, but here *prices* are being depicted. What is there in a barrel which contains a price – a price tag? The price is around 5.5 times higher in 1979 than in 1973, but the barrel's volume is 500 times larger!

Chart B: A peculiar road!

The picture attracts the attention of those interested in cars. The reader imagines sitting in a car and seeing a road ahead. The picture was published in 1978 and was depicting the future.

The picture is suggestive, but the data is illustrated very badly. The time axis is strange because the most distant year is closest to the reader and the current year is far away. The lengths of the lines across the road give very misleading size comparisons. Tufte's comments on the diagram indicate this. The picture is also slanting the wrong way – in the U.S.A. people drive on the right.

This line, representing 10 miles per gallon in 1978, is 0.6 inches long.

B

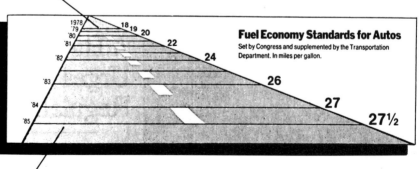

This line, representing 27.5 miles per gallon in 1985, is 5.3 inches long.

New York Times, August 9, 1978, p. D-2.

Chart C: Spot on!

The picture attracts attention, has a well-founded association with the subject and also gives a good picture of the data since the heights of the bars give an approximate picture of the size of imports. The three-dimensional bars give a natural impression since they are included in the picture as ship's cargo.

Chapter 11 discusses bar charts with a false third dimension. By rotating the axes we get charts of type D, where the third dimension does not have any function. We see that, in principle, the oil tanker's cargo is such a bar chart, but here the third dimension has an important function to fulfil – to attract attention and provide associations with the subject.

Symbols or words?

Barometer charts E and F show that the explanatory text in E can be replaced by easily understood symbols to good effect. The reader can then grasp the contents more quickly.

Discreet backgrounds

Chart G is a normal bar chart which has been given a background. This is a discreet picture of a farm which gives a suitable link to the contents.

With today's PC technology it is easy to include symbols and background pictures in charts.

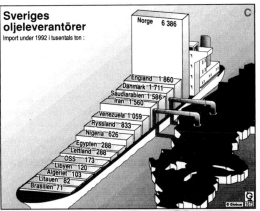

Sveriges oljeleverantörer
Import under 1992 i tusentals ton :

The oil comes from Norway

Last year Sweden bought 17 million tonnes of crude oil. By far the largest supplier was our neighbouring country in the west, Norway, which accounted for over six million tonnes. In second place came the U.K., in third Denmark and only in fourth place was the first OPEC country – Saudi Arabia.

Source: Nerikes Allehanda, March 30, 1993

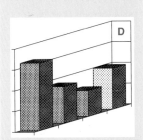

This chart is discussed on page 71

Energy consumption per person-kilometre
Kilowatt-hours

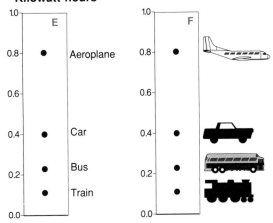

The symbols in chart F and the background in chart G are taken from picture libraries for PC (clip art)

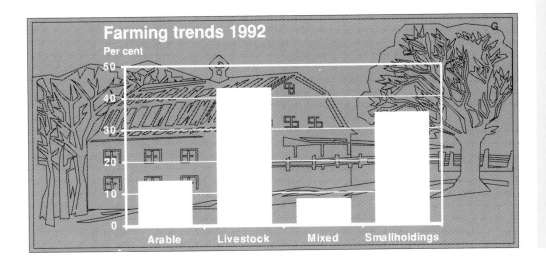

14 Check-list

Before you start

What we say about charts in this chapter is also true for maps and pictures

What is my target group?

What is a good chart for one category of readers can be far too difficult for another. The basic principle is of course that better qualified readers can cope with more difficult charts and that they do not need charts which only alleviate the text. Remember not to underestimate the capacity of the reader to understand even difficult charts. One of the reasons why we draw charts is that we want to depict complicated relationships.

What is the role of the chart?

Be quite clear about what role the chart is to play: to capture the attention of the reader on the cover of a publication, to lighten up a heavy piece of text, to show a large set of data or to show relational or developmental patterns. Different roles lead to different approaches.

What kind of chart should I choose?

The choice of chart type depends on the subject matter, the target group and the set of data. The subject matter determines what comparisons are important. The target group determines how complicated the chart can be. The characteristics of the variables determine, for example, whether you should use a bar chart or a histogram.

How should the chart be presented?

Different presentational methods and technologies have different requirements. An obvious question is whether you have access to colour or high-quality printing, which give greater scope for expression. A chart on an overhead projector can be made quite complicated, because the person who presents can give explanations.

How big should the chart be?

Obviously you are able to work with more details if the chart covers half a page than if it covers a quarter of a column. You should get a clear idea from the start (perhaps with the help of a synopsis) how big the chart will be in print.

Is a *chart* really best?

In some situations a table or possibly even a descriptive text is more suitable. This is true, for example, in situations where the differences between groups or points in time are very small and a chart would be uninteresting, and in situations where small differences are important and a chart would often be unclear.

Is *one* chart really best?

In many situations several simple charts are better than one complicated one. For example, it is often difficult to show more than two groups in one chart satisfactorily. Do not forget the possibility of using systems of small charts with the same basic appearance.

What technology should I choose?

It is best if you can exploit the advantages of modern PC technology. Then it is easy for you to try out various alternatives and the whole time you can see how the chart suits the actual data. In most cases charts drawn on computer can be used directly as originals for printing. If you are drawing by hand you can make rough sketches to compare different chart types and then allow an expert to make the final original.

When you (think you) have finished

Is the chart easy to read?

What you want to show in the chart has to become evident quickly and clearly. Check that the title is simple and easy to understand, but still complete. A suitable number of grid lines make it easier for the eye to orientate itself. Good patterns make it easier to read and understand the chart. In this respect it is important to remember the target group for which you are drawing the chart.

Can the chart be misinterpreted?

Does the chart really show what you had in mind? Make sure that the scales cannot give rise to misunderstandings. It must be clear what kinds of scales and definitions are being used and what they mean. Remember to choose the same scale if several charts are to be compared.

Does the chart have a good size and shape?

Have you followed the column width for the size of the chart? The chart should extend over one or two columns depending on what is to be shown. If there is little information in the chart then it is a good idea to make a smaller chart which will fit into one column.

   If you have several charts one above (or next to) the other, it may be appropriate to make them the same width (height).

Is the chart in the right place?

The chart should always be placed close to the text where the topic is dealt with. The positioning on the page is also important. Successful positioning on the double-spread of pages creates a neat and harmonious layout.

   If you have several charts one above the other then either the y-axes should be in line or the charts should be displaced neatly sideways. Similarly, it may be appropriate for charts which lie side by side on the same double-spread of pages to have their x-axes on the same line.

Does the chart benefit from being in colour?

In an expensive publication you may choose to use colour, but the colour should then have a purpose and not simply function as a decoration. Giving a chart, map or other picture colour is a very simple operation with the majority of computer layout programs. However, originals of colour charts take somewhat longer to produce and the printing process is more expensive.

Have you tried the chart out on anybody?

Test the chart out on somebody who can be considered to correspond to the target group before you make the final diagram. Ask questions about the chart so that you understand how the subject of the test perceives your chart.

> **A poor chart is worse than no chart at all!**

15 Choice of technology for producing charts

The basic principle for designing and drawing good charts is to feel your way towards the chart which fulfils the objective in the best way possible. When the chart type is chosen on the basis of the characteristics of the set of data it is appropriate to make a number of chart sketches in order to get an idea of how the chart should ultimately look. Both the sketches and the final chart can be done by hand or on computer.

Drawing by hand versus computer

A number of years ago all chart drawing was done using the ability of the hand to use pen and paper together with aids such as rulers and stencils etc. Charts are still drawn by hand, but now by tapping on keys and working with a mouse, light pen or similar tools.

Thus drawing by hand has increasingly disappeared in favor of ADP support in the production of charts. Up until the mid 1980s most chart drawing was carried out on mainframes. Now almost all production of charts is carried out using PC or Macintosh programs. Irrespective of the technology used all charts drawn on computers should be designed well so that the final result is correct and easy to interpret.

In his or her enthusiasm for the possibilities offered by the new technology it sometimes happens that the designer tries to convey far too much in one go. The result is usually charts which are overburdened and cluttered with too many ill-chosen patterns and colours. It is also easy to wallow in amusing chart types, such as pictograms and three dimensional charts. Always bear in mind that the technology should be subordinate to the statistical message.

Do not forget that it takes time and requires a great deal of training to produce neat and statistically correct charts in a good and efficient way

Computer technology

If you have decided to use ADP technology for the production of charts it is important to choose from the great variety of computers, software programs and printers (drawing equipment) which is on the market. Different computer programs can handle different types of chart, and printers produce different qualities of patterns and lines. The thing here is to make the right choice for your particular requirements.

Software

There are many programs to choose between for drawing charts and maps on PCs and Macintoshes. Development is rapid and new programs are being introduced onto the market all of the time. New versions of already established programs also appear at regular intervals. When you choose a program for a PC it is best to choose one which is made for the Windows environment. Programs under Windows work together better than programs outside Windows.

In order to produce publications which contain text, charts, maps and other pictures, desk top publishing programs are used. In these programs documents are put together by importing the different parts from statistical programs, table programs, word processing programs, charting programs, mapping programs, and programs for scanning pictures.

All graphics programs basically consist of instructions for producing and manipulating dots. The dots are used to create so-called primitives such as lines, areas and texts. The primitives are then put together to form different types of chart

Printers

There are a number of printers to choose between, everything from simple dot matrix printers to advanced photocompositors. The most common at the moment (1994) are laser printers with resolutions of between 300 and 600 dots per inch. The choice of printer, and the choice of operating mode, determines how the chart will turn out. The neater you want your charts to be, the higher the resolution you need to choose on your laser printer. Most laser printers on the market can only print out charts in black and white. Colour laser printers have been introduced, though, and will probably come to predominate in future.

PC technology in the production of this book

Graphing Statistics & Data was produced entirely in a PC environment. The user interface was Windows 3.1 with accompanying programs, operating modes and printers. On this page we give an overview of which programs were used to produce this book.

Data files
The data for all of the charts and maps which have full titles is based on data from Statistics Sweden's publications.

Charting programs
Charisma 2.1 was used for making most of the charts and some of the simpler maps. Some of the charts were created using Excel 4.0.

Mapping programs
MapMaker 1.10 was used to produce the more complicated maps in chapter 10.

Pictures
Some of the pictures in the book were photographed (copy proof), such as Napoleon's campaign on p. 9, or taken from clip art libraries, such as the aeroplane on p. 87.

Graphics files
The book contains about 300 charts and maps which were created in the program.

Desk top publishing programs
PageMaker 4.0 was used as the desk top publishing program. The text was written directly in PageMaker and the graphics files were imported into the document. When they were imported links were created to the directories with the graphics files.

Document files
A number of document files were created in PageMaker, one for each chapter.

Printers
For test print-outs we used a laser printer with a resolution of 300 dpi. To print the originals we used a photocompositor with a resolution of 1270 dpi.

Computer technology is changing all the time. The technology available to-day (1994) will be out-dated tomorrow. A few years ago it would not have been possible to produce this book with the technology we have used

The alternative to photographing pictures or taking them from picture libraries is to use a scanner. A scanner reads a picture into a file which can then be imported into a DTP program

data files   data files

Charisma

MapMaker

charts   maps   pictures

PageMaker

documents

Laser printers   Photocompositors

Almost all of the charts in the book can be created with your own graphics program. It is your knowledge and creativity which govern what you can do with the program. Do not give up if you do not see all of the possibilities immediately, instead, try things out and keep battling until the problem is solved.

Index

*Paging in **bold** indicates the most important page*